The Internet as a Large-Scale Complex System

Santa Fe Institute
Studies in the Sciences of Complexity

The Internet as a Large-Scale Complex System

Editors

Kihong Park
Purdue University

Walter Willinger
AT&T Labs—Research

Santa Fe Institute
Studies in the Sciences of Complexity

OXFORD
UNIVERSITY PRESS
2005

OXFORD
UNIVERSITY PRESS

Oxford University Press, Inc., publishes works that further
Oxford University's objective of excellence
in research, scholarship, and education.

Oxford   New York
Auckland   Cape Town   Dar es Salaam   Hong Kong   Karachi
Kuala Lumpur   Madrid   Melbourne   Mexico City   Nairobi
New Delhi   Shanghai   Taipei   Toronto

With offices in
Argentina   Austria   Brazil   Chile   Czech Republic   France   Greece
Guatemala   Hungary   Italy   Japan   Poland   Portugal   Singapore
South Korea   Switzerland   Thailand   Turkey   Ukraine   Vietnam

Copyright © 2005 by Oxford University Press, Inc.

Published by Oxford University Press, Inc.
198 Madison Avenue, New York, New York 10016

www.oup.com

Oxford is a registered trademark of Oxford University Press

Library of Congress Cataloging-in-Publication Data
The Internet as a large-scale complex system / editors,
Kihong Park, Walter Willinger.
p. cm. — (Santa Fe Institute studies in the sciences of complexity)
Includes bibliographical references and index.
ISBN-13 978-0-19-515720-8; 978-0-19-515721-5 (pbk.)
ISBN 0-19-515720-6; 0-19-515721-4 (pbk.)
1. Internet. 2. System analysis. 3. Telecommunication—Traffic.
I. Park, Kihong, 1964– II. Willinger, Walter, 1956– III. Proceedings volume
in the Santa Fe Institute studies in the sciences of complexity.
TK5105.875.I57I5353 2005
004.67'8—dc22   2004063562

9 8 7 6 5 4 3 2 1

Printed in the United States of America
on acid-free paper

About the Santa Fe Institute

The *Santa Fe Institute* (SFI) is a private, independent, multidisciplinary research and education center, founded in 1984. Since its founding, SFI has devoted itself to creating a new kind of scientific research community, pursuing emerging science. Operating as a small, visiting institution, SFI seeks to catalyze new collaborative, multidisciplinary projects that break down the barriers between the traditional disciplines, to spread its ideas and methodologies to other individuals, and to encourage the practical applications of its results.

All titles from the *Santa Fe Institute Studies in the Sciences of Complexity* series carry this imprint which is based on a Mimbres pottery design (circa A.D. 950–1150), drawn by Betsy Jones. The design was selected because the radiating feathers are evocative of the outreach of the Santa Fe Institute Program to many disciplines and institutions.

Contributors List

David Ackley, *University of New Mexico, Computer Science Department, Albuquerque, NM 87131; e-mail: ackley@cs.unm.edu*

Justin Balthrop, *University of New Mexico, Computer Science Department, Albuquerque, NM 87131; e-mail: judd@cs.unm.edu*

Jie Chen, *School of Electrical and Computer Engineering, Phillips Hall, Cornell University, Ithaca NY 14853; e-mail: jchen@ece.cornell.edu*

Stephanie Forrest, *Department of Computer Science, University of New Mexico, Farris Engineering Center, Albuquerque, NM 87131; e-mail: forrest@cs.unm.edu*

Matthew Glickman, *University of New Mexico, Computer Science Department, Albuquerque, NM 87131; e-mail: glickman@cs.unm.edu*

Ramesh Govindan, *University of Southern California, Computer Science Department, 941 W. 37th Place, Los Angeles, CA 90089-0781; e-mail: ramesh@usc.edu*

Matthias Grossglauser, *School of Computer and Communication Sciences (I&C), Room INN-040, LCA/I&C, EPFL, Lausanne CH-1015, Switzerland; e-mail: matthias.grossglauser@epfl.ch*

Tomoko Itao, *Network Intelligence Laboratory, NTT Network Innovation Laboratories, Nippon Telegraph and Telephone Corporation (NTT), 9-11 Midori-Cho 3 Chome, Musashino-shi Tokyo, 180-8585, Japan; e-mail: tomoko@ma.onlab.ntt.co.jp*

H. T. Kung, *33 Oxford Street, Cambridge, MA 02138; e-mail: kung@harvard.edu*

Jörg Liebeherr, *Computer Science Department, University of Virginia, 151 Engineer's Way, Charlottesville, VA 22904-4740; e-mail: jorg@cs.virginia.edu*

Masato Matsuo, *Network Intelligence Laboratory, NTT Network Innovation, Laboratories Nippon Telegraph and Telephone Corporation (NTT), 9-11 Midori-Cho 3 Chome, Musashino-shi Tokyo, 180-8585, Japan; e-mail: matsuo@ma.onlab.ntt.co.jp*

Kihong Park, *Department of Computer Sciences, Purdue University, West Lafayette, IN 47907; e-mail: park@cs.purdue.edu*

Jennifer Rexford, *AT&T Labs—Research, 180 Park Avenue, Room A-139, Florham Park, NJ 07932; e-mail: jrex@research.att.com*

Scott Shenker, *The ICSI Center for Internet Research (ICIR), 1947 Center Street, Suite 600, Berkeley, CA 94704; e-mail: shenker@icir.org*

Tatsuya Suda, *Information and Computer Science, University of California-Irvine, Irvine, CA 92697-3425; e-mail: suda@ics.uci.edu*

Hongsuda Tangmunarunkit, *Bank of Thailand, Chor Building, 273 Samsen Rd., Bangkhunphrom, Bankok 10200, Thailand; e-mail: hongsudat@bot.or.th*

James S. Thorp, *Virginia Tech, 340 Whittemore Hall, Blocksburg, VA 24061; e-mail: jsthorp@cvt.edu*

Walter Willinger, *AT&T Labs—Research, 180 Park Avenue, Florham Park, NJ 07932; e-mail: wllter@research.att.com*

C. H. Wu, *Institute of Information Science Academia Sinica, Taiwan*

Contents

Preface

The Internet may be viewed as a "complex system" with diverse features and many components that can give rise to unexpected emergent phenomena with implications to understanding and engineering the Internet. This book brings together chapter contributions from the "The Internet as a Large-Scale Complex System" workshop held at the Santa Fe Institute (SFI) in March 2001, jointly sponsored by the National Science Foundation (NSF; grant ANI-0102129) and SFI.

The objective of the workshop—and this book—was to capture a snapshot of features of the Internet that may be fruitfully approached using a complex systems perspective, meaning the interdisciplinary nature of tools and methods that have been used to tackle various complex systems. With the Internet penetrating the socioeconomic fabric of everyday life, a broader as well as a deeper grasp of the Internet may be needed to meet the challenges facing the future. They include: (1) a demand for greater service functionality, in particular, quality of service (QoS), robustness, and security; (2) traffic burstiness, heavy-tailed workloads, and power-law network connectivity which impact resource provisioning

The Internet as a Large-Scale Complex System,
edited by Kihong Park and Walter Willinger, Oxford University Press.

and traffic control; and (3) scalability of the underlying networking infrastructure to support orders of magnitude more users, hosts, and servers. Scalability looms as an imposing problem, not only to the control algorithms of the past, but also for the provisioning of user-specified QoS and effective network security which, in turn, may require the interaction—not necessarily cooperative—of different networking elements with possibly uncertain consequences.

The dual aim of this book is to provide an exposition to topics relevant to the "Internet as a Complex System" theme and a first encounter with networking methodologies that can help provide part of the grounding needed in technical discussions of the Internet. The first chapter, "The Internet as a Complex System," gives an overview of a complex systems approach to studying the Internet and discusses three specific examples—self-similar network traffic, power-law network connectivity, and noncooperative network games—where further integration and assimilation under a complex systems umbrella may be interesting to explore. The second chapter, "Passive Traffic Measurement for Internet Protocol Operations," discusses the problem of inferring traffic congestion in Internet Protocol (IP) internetworks using traditional techniques from network management and more recent developments from network tomography. The third chapter, "Internet Topology: Discovery and Policy Impact," presents, in two parts, the challenges associated with Internet connectivity discovery, a solution based on Mercator, and a discussion of how policy considerations can influence routing performance. Chapters 2 and 3 convey the importance of measuring the Internet, a trend in network research methodology prompted, in large measure, by the discovery of self-similar traffic in the early 1990s. Akin to the interplay between theoretical and experimental physics critical to its grounded development, recent networking research has emulated the give-and-take with successful results in the areas of self-similar traffic and power-law topology.

Chapter 4, titled "Hidden Failures: The Role of the Protection System in Major Disturbances in Power Systems," gives an overview of cascading failures in power grids that have resulted in large-scale electrical blackouts. Some parallels to Internet fault-tolerance may be drawn, especially with respect to the clustered occurrence of "bad" events that violate independent failure models due to correlation-at-a-distance created by domino effects. In Chapter 5, "A Note on Statistical Multiplexing and Scheduling in Video Networks at High Data Rates," an introduction to statistical multiplexing and scheduling is given that harnesses the law of large numbers availed by the large-scale nature of the Internet to facilitate efficient traffic control. In Chapter 6, "Content Networks: Taxonomy and New Approaches," an overview of content networks is provided that shows how information on the Internet may be organized to provide content-driven access at the routing level through "semantic nets." The organizing principles also have relevance to caching and World Wide Web information retrieval.

Chapter 7, "Computation in the Wild," presents a biologically motivated system for effecting network intrusion detection called LISYS. In Chapter 8, titled "The Bio-Networking Architecture: A Biologically Inspired Approach to the

Design of Scalable, Adaptive, and Survivable/Available Network Applications," analogies from biology such as self-organization, adaptation, and evolution are employed to create autonomous distributed network applications. Chapters 7 and 8 provide application-oriented agent network environments that embody biologically inspired principles aimed at facilitating adaptation, robustness, and scalability.

The material covered in this book is but a first step toward exploring a complex systems approach to understanding the structure and dynamics of the global Internet. We hope that this book can serve as a useful reference for future efforts in this direction. Before concluding, we would like to acknowledge the help and support from SFI, without whose active backing and resources the book and workshop would not have been possible. Special thanks to Andi Sutherland and Kevin Drennan for their organizational assistance with the workshop. We would like to extend our gratitude to Ronda Butler-Villa, Della Ulibarri, and Laura Ware for their help with putting together this book. Della Ulibarri's tireless efforts have been instrumental in bringing this book to fruition. Lastly, we would like to thank Karen Sollins for her encouragement and support of the workshop during her tenure as program manager at NSF.

Kihong Park   Walter Willinger
Purdue University   AT&T Labs-Research

The Internet as a Complex System

Kihong Park

1 INTRODUCTION

The Internet, defined as the worldwide collection of Internet Protocol (IP) speaking networks, is a multifaceted system with many components and diverse features. In recent years, it has become an integral part of the socioeconomic fabric. Although the Internet is still an evolving system and, therefore, a moving target, understanding its properties is relevant for both engineering and potentially more fundamental purposes. The Internet is a complex system in the sense of a complicated system exhibiting simple behavior, as opposed to a simple system exhibiting complex behavior. The latter aspect forms the cornerstone of studies of complex systems whose rigorous underpinning is provided by dynamical systems theory. A canonical example is the logistic equation from population dynamics which possesses a rich structure including chaotic orbits and fractal non-divergent domains [41]. A somewhat different example of simple systems capable of generating complicated behavior are many-body systems, in particular, interacting particle systems [82]. We would classify such systems, despite the

The Internet as a Large-Scale Complex System,
edited by Kihong Park and Walter Willinger, Oxford University Press.

large number of components they admit, as simple due to the homogeneity and limited capability of each component—in Ising spin systems each component has two states—and the local interaction allowed to the components. Statistical mechanics studies macroscopic and emergent properties of such systems, with ergodic theory and percolation theory providing part of the mathematical foundation. Cellular automata (discrete-time interacting particle systems) are capable of universal computation which gives rise to undecidability problems and computation theoretic considerations of dynamical systems.

We view the Internet as an instance of a complicated system exhibiting simple behavior due to a compendium of innate architectural features that span a range of scientific disciplines. They include:

- the characteristics of network traffic when viewed as a time series;
- closed-loop and open-loop control governing the flow of traffic which involves control theory;
- the connectivity structure of information networks which engages graph theory;
- the behavior of users and protocols—protocols are formal rules and conventions underlying information exchange—in resource-bounded, competitive environments which involves game theory; and
- the organizational behavior of Internet Service Providers (ISPs), which influences the network infrastructure, including peering relationships among ISPs, a domain of social sciences.

A discussion of these features is the subject matter of this article.

There are few established examples of complex systems exhibiting simple behavior, excluding tautological cases such as the natural phenomena and systems studied in physics and biology. The Internet is comparatively more transparent and tractable, perhaps requiring on the order of decades—as opposed to centuries for physics and biology—to uncover and understand its workings. At its physical basis, the Internet is a flow network in which information transmission is governed by communication theory pioneered by Shannon [125]. Quantum communication [14, 15], which uses entanglement in quantum mechanics for efficient communication, is in its early stages, too early to be included in the present discussion of the Internet. As a complex system exhibiting simple behavior, what makes the Internet unique are its two distinguishing characteristics: (1) it is a melting pot where the key ingredients represent a confluence of several disciplines, and (2) the Internet is perhaps the largest man-made many-body system. Jointly, the interdisciplinary nature of the architectural features and the engineered nature of the many-body system give rise to novel phenomena, modeling challenges, and synergistic opportunities, which have a chance of being scientifically grounded and form another cornerstone of complex systems.

An example of the synergistic opportunity is the enhanced effectiveness of game theory when played out in the context of the Internet. Von Neumann and

Morgenstern advanced game theory [138], in part, to establish a mathematical foundation for economics—referred to, then, by many as the dismal science. In some respects, economics, notwithstanding the wealth of beautiful mathematics and modeling work carried out since then, has remained a dismal science for the simple reason that it continues to lack sufficient predictability. Econometric models cannot adequately account for the effect of political upheaval—never mind predicting their occurrence—a consequence of the Achilles' heel of social sciences: limited ability to factor the dynamics of human behavior, collective or singular. Game theory can provide important qualitative insights but ultimately suffers under the same predicament. A competitive game involving several human participants may be modeled by a corresponding noncooperative game involving rational self-optimizing decision processes; however, the outcomes may agree or may not agree—it all depends on what the human players actually decide to do.

The situation is less bleak when game theory is fused with the Internet. The Internet injects a measure of predictability—through behavior codification—effected by protocols that sit between a human user and the communication medium handling the transfer of information. The bulk of Internet traffic is comprised of file transfers that arise from Hypertext Transfer Protocol (HTTP) based Web traffic which, in turn, is governed by Transmission Control Protocol (TCP). TCP is a cooperative protocol that tries to achieve speedy, reliable communication. "Cooperative" means that TCP behaves in a gentlemanly manner upon detecting possible congestion by throttling the traffic submitted to the network. The behavior of the TCP protocol is standardized by the Internet Engineering Task Force (IETF), the standards body of Internet protocols. Protocols, acting as automated assistants embodying codified behavior, induce a well-behaved environment without constant subjection to the inner workings and whims of human decision making. Questions involving stability and efficiency can be addressed, and quantitative predictions advanced with scientific certainty. That is not to say that humans play no part. The selection and timing of information transfer requests is under the control of human users, which puts us in the driver's seat. However, the time scale of information transmission and control in broadband networks is at the millisecond level and below, a regime where most humans would be out of place. Thus, the total picture is that of a two-layered playing field separated by time scale where the influence of human decision making is partly delimited and isolated. Whether this crack provides a sufficient opening for reigning in the unwieldy influence of human decision making remains to be seen. The potential is there, making the Internet a fertile playing ground for a more effective game theory.

The preceding example is but one of several that will be discussed in this chapter. In the next section, we give a bird's-eye view of five key architectural features of the Internet—self-similar traffic, scalable traffic control, power-law connectivity, game theory and quality of service (QoS), and organizational behavior—which are related to the theme, the Internet as a complex system. This is followed by sections discussing the known aspects, consequences, and challenges. The selected features meet two criteria: networking relevance and technical novelty. Networking relevance means that the complex system trait has direct bearing on Internet traffic engineering and is not a mere curiosity. Technical novelty means that understanding the complex system trait involves more than straightforward applications of known ideas and techniques.

2   INTERDISCIPLINARY FEATURES OF THE INTERNET

2.1   A PACKET'S JOURNEY

We motivate the selection of the interdisciplinary architectural features by describing a typical—and on the surface mundane—sequence of events that transpires on the Internet. We will use this example as a skeleton to attach the five features which will, then, be given concrete meaning. The description, although oversimplified, may help give the general reader a logical glimpse of network mechanisms and their intricacies, in addition to introducing needed terminology.

Suppose a user runs a Web browser at an end system—PC, laptop, or handheld device—and clicks on a link containing the location information of an object such as a Hypertext Markup Language (HTML) document or some other file (e.g., executable binary, audio or video data) that is to be accessed using HTTP. This triggers an HTTP request message that is passed down to TCP in the protocol stack—protocols in the operating system (OS) of an end system are organized in a partial order represented as a protocol graph—which encapsulates the HTTP request by treating it as payload. This is akin to an already sealed letter going into a FedEx envelope. TCP memorizes the packet information in the event it needs to resend it: the Internet is "leaky." TCP's packet is handed down to IP, a protocol responsible for routing. IP determines where to forward the packet so that it can come closer to reaching its final destination. IP performs its own encapsulation and hands the resultant packet to the link layer. A popular link layer is Ethernet—standardized by Institute of Electrical and Electronics Engineers (IEEE) under IEEE 802.3 for wired and IEEE 802.11 for wireless media. The link layer has access to the physical address of the next hop's IP address—every network device has a unique physical address—encapsulates the IP packet with an envelope containing physical addresses, and hands it down to the physical layer. The physical layer oversees the transmission of information containing the link layer packet over its communication medium.

The physical layer at the receiving end decodes the transmission and does a hand-off to the appropriate link layer protocol above. Assuming the receiver is an IP speaking device, the link layer protocol decapsulates and hands it off to the IP layer, which determines whether additional forwarding is required to reach the final destination. If so, the packet is encapsulated and passed down the protocol stack. This process is repeated at every IP-enabled device—called router—on the forwarding path until the destination IP device is reached. At that point, the IP layer passes its payload up to TCP which, in turn, hands off its payload to HTTP, and HTTP to its application. In this example, it is a Web server that processes the HTTP request. This prompts an HTTP response, which is passed down the protocol stack at the destination IP device and returned to the original sender, the client. Several things can go wrong on an IP packet's journey. The packet, during physical transmission, may get corrupted due to noise or interference—especially severe in wireless segments of the Internet—which results in dropping the corrupted packet when so detected. The packet, upon arriving at a router, may find the router busy processing other waiting packets, and even worse, find no room or buffer space, which causes the packet to be discarded. Less frequently, a router may fail, erasing the transiting packet residing in its memory. IP has no provision for dealing with packet loss, which means that TCP running at the sender with the assistance of its counterpart running at the receiver is the earliest point at which recovery can be attempted. The dumb network core/intelligent network edge is characteristic of the Internet's design, referred to as the end-to-end paradigm [33].

2.2   COMPLEX SYSTEMS FEATURES

We revisit the packet journey example and point out several innate, interdisciplinary features that reside with the skeleton of packet forwarding and information transmission mechanics.

Self-similar Traffic.   An important engineering consideration is the load or traffic—measured in bits per unit of time—impinging on a bottleneck router, as excessive traffic can result in congestion and information loss. The same is true of end systems, in particular, servers. If traffic, viewed as a time series, is undulating with severe peaks and valleys, the capacity allocated to handle demand must be correspondingly bursty to match the time-varying demand. This is to avoid losses—translated to delay when sufficient buffer space is available to hold excess traffic—and reduce resource wastage stemming from overprovisioning, which carries an economic cost. Flat traffic is desirable because it is predictable and obviates the need to frequently reshuffle resources which, in many instances, is difficult to do. Whatever our wishes, Internet traffic may follow its own set course, and understanding its properties is fundamental to effective traffic engineering. Two obvious factors that influence traffic demand are the arrival pattern, in time, of user requests—that is, clicking on a Web link in the packet journey example—and the size of the file or information object requested. Other things being equal, the more frequent the user requests and the larger the requested files, the higher the average load experienced by a network system. Our interest concerns the shape of the resultant traffic.

From a server's perspective or a router's perspective that lies on frequented paths to popular servers, request arrivals from different users may be construed to be independent, at least at time scales on the order of minutes and below. If indeed so, the law of large numbers (LLN), assuming users are many, induces statistical regularity conducive to flatness on two fronts: one, the number of user requests occurring during a time window will concentrate around a mean, and two, the number of user requests across two disjoint time windows will become uncorrelated. To a first approximation, the arrival interval between successive requests has been observed to be independent, with an exponential distribution and resultant total load that is Poisson. The action of LLN across space and time has been the central property targeted and harnessed by traffic engineering tools in telephony and data communication. To the surprise of many, traffic measurement collected in the late 1980s at Bellcore on Ethernet showed that traffic was bursty at large time scales [81], inconsistent with the flatness predicted by independent or weakly correlated arrivals. In fact, Bellcore's Ethernet data exhibited self-similarity in the sense of variability, as captured by correlation, remaining invariant across a wide range of time scales: from milliseconds to seconds to tens of minutes. This phenomenon has been confirmed in other contexts since then and shown to be the norm rather than the exception. One factor in the packet journey example we ignored is the size of requested files. It turns out that size does matter, and the peculiar distribution of file sizes (most files are small but a few that are very large[1] are also referred to as "mice and elephants") affects the fractal characteristic of Internet traffic. Traffic self-similarity is an emergent, macroscopic trait of the Internet whose seed can be found in microscopic properties of its components. The fractal dynamics of Internet traffic make it a representative example of the Internet as a Complex System metaphor.

Scalable Traffic Control. The self-similar nature of Internet traffic has served to draw attention to the importance of empirical measurement, almost to the extent of treating the Internet as a "natural" phenomenon in its own right, the properties of which, even to the designers, may initially remain hidden. The Internet, notwithstanding its phenomenological richness, is an artificial, engineered system, a key distinguishing feature compared to complex systems arising in nature. The Internet is a controlled system where the bulk of daily traffic is regulated by TCP, a feedback control designed to effect speedy, reliable data transfer while avoiding congestion. TCP implements nonlinear control, is subject to the effect of feedback latency, and is instantiated in an environment containing thousands of other TCP flows competing for shared network resources—bandwidth and buffer space—at bottleneck routers. For example, the HTTP session in the packet journey example is but one of many traversing a bottleneck link. Coupled TCP flows, when conditions are ripe, may synchronize [145]—a phenomenon abundant in nature such as the synchronized flashing of fireflies [112]—leading to periodic underutilization and overutilization. Even without oscillatory synchronization, TCP is subject to stability problems due to feedback, latency and congestion avoidance. Fairness—in the sense of equal share—may be violated when one TCP flow sharing a bottleneck link with another has to traverse many

---

[1]The technical definition, "heavy-tailedness," will be discussed in section 3.

more hops, which puts the longer flow at a comparative disadvantage: due to increased feedback latency, news arrives more slowly and consequently is less useful in control actions. TCP, in some instances, can contribute to self-similarity of network traffic as its nonlinear control, when reducing the sending rate during persistent packet loss, injects exponentially increasing wait periods that can translate to prolonged idleness, a form of correlation. However, it is a secondary factor compared to the dominant role played by file size distribution. Congestion control, of which TCP is an instance, has the potential to induce complicated dynamics, representing another aspect of complex systems.

Related observations hold for routing, the second of the two major Internet traffic controls. Synchronization of routing updates can affect cyclic busy periods during which part of a router's capacity is turned away from packet forwarding [54]. Routing tables may take a long time to stabilize, creating time windows during which packets are misrouted [80]. Quality of service (QoS) is another important pillar of traffic control. Its present impact, however, is limited compared to congestion control and routing due to almost nonexistent deployment. That is not to say that routers on the Internet do not possess QoS capabilities. They, in fact, are endowed with a slew of IETF standardized mechanisms, but these features are not activated outside of isolated regions. Several of these are candidate building blocks in a future QoS-enabled Internet, still a technical as well as a socioeconomic challenge.

Power-law Connectivity. In the late 1990s, two separate, but intimately related, phenomenological features of the Internet were discovered [4, 50]. The first concerns the connectivity structure of the World Wide Web (WWW), where two Web pages, viewed as nodes in a graph, are defined to be joined by an edge if there is a link from one to the other. The second concerns the business-to-business relationship between domains or autonomous systems that represent organizational units. Autonomous system (AS) is a technical term and part of the Internet standard. The Internet AS topology is defined as a graph where the nodes are ASes, and an edge exists between two nodes if the ASes—more precisely, border routers belonging to the ASes that affect the transfer of packets between domains—are connected by one or more physical links representing customer-provider relationships. What these measurement graphs showed is that their connectivity structure is far from random, exhibiting power-law decay—as opposed to exponential decay—in the size of the neighborhood of a node. Random means that a node is connected to any other node with some fixed, independent probability. In the case when the probability is 1/2, this is tantamount to saying the set of all graphs with a given number of nodes, when viewed as a discrete sample space, has uniform probability. The measurements showed that not all nodes are created equal, and not all graphs are created equally likely.

A characteristic trait of random graphs is that most nodes look alike, possessing about the same number of links that are strewn all over the place. The probability that a node has a neighborhood size deviating from the mean is exponentially small, a consequence of LLN. In WWW and Internet domain graphs, the decay was observed to be polynomially small—a power law, and thus the name power-law graph—admitting nodes with very many neighbors. With 20/20 hindsight, it is perhaps not unusual that WWW and Internet domain graphs should exhibit power-law connectivity. During long flights one may have browsed through an airline magazine, finding the carrier's route map in the back pages. Power-law graphs, visually, resemble airline route maps where nodes corresponding to major hubs have many links that connect smaller regional airports. Some regional nodes connect to more than one hub, and hubs are connected to each other through a backbone. Two important aspects of power-law graphs are captured by the expressions: "the rich get richer and the poor get poorer," and "a few are connected to many, many are connected to a few." The first represents a dynamic viewpoint, whereas the second conveys a static structural aspect. A number of studies in the 1950s [86, 126, 146] showed power-law (also called Zipf's law in special cases) skews in social phenomena which were attributed to the intuition that already popular entities were likely to become even more popular—in part, through a bandwagon effect—whereas unpopular entities faced the opposite predicament. In Internet domain graphs, a major service provider acting as a conduit for other providers and stub customers—a stub AS is a domain that does not provide transit service to other domains—is likely to attract more customers compared to smaller providers. New customers may perceive an advantage in connecting to an established provider with an existing, large customer base that is accessible by a single hop. This is in addition to the perceived reliability associated with established organizations and brand names. An extreme form of a few nodes being connected to many and many being connected to a few is the star topology: there is a single central node from which all other nodes emanate. When there are multiple high degree nodes—the degree of a node is the number of links incident on the node—they resemble locally star-like subgraphs that are connected to each other through a backbone network. There are a number of consequences of power-law connectivity for network security and performance, some relating to percolation and phase transition. They will be discussed in section 4.

Game Theory and QoS.   As indicated in the Introduction, game theoretic considerations arise naturally in congestion control where multiple sessions share a common bottleneck link through which traffic must be scheduled. In the packet journey example, two servers transmitting information to clients across a common bottleneck may distinguish themselves by employing clever congestion controls that outperform the competition—inclusive throughput gains obtained at the expense of others—that translate to faster response times and commercial advantage. Supposing throughput is the performance metric that selfish users and their protocols aim to maximize, the noncooperative congestion control game that pits selfish congestion control protocols against each other is an instance of Tucker's Prisoner's Dilemma game. In the two-player setting, at the onset of congestion, the protocols can act cooperatively and throttle their sending rate, alleviating congestion and attaining an equitable share of the total achievable throughput. This corresponds to neither prisoner, when interrogated in separate rooms, ratting out the other. Each prisoner receives a two-year sentence with possibility of parole. If one protocol acts cooperatively but the other does not—the cooperative protocol backs off when congestion arises while the noncooperative one fills the slack—the game leads to a state where bandwidth is monopolized by the noncooperative protocol. This corresponds to the cooperative prisoner receiving a 20-year sentence for staying mum, while the selfish prisoner who betrayed his colleague gets off scot-free. When both protocols behave selfishly, congestion persists with each achieving a small throughput.

The Prisoner's Dilemma illustrates that noncooperative games, in general, lead to equilibria—if they exist—that are less desirable than those reachable by corresponding games where players are cooperative. Indeed, only under special circumstances does Adam Smith's invisible hand lead to an efficient orchestration of shared resources. A somewhat different example of the adverse influence of selfishness is Braess's paradox [20]. It describes a situation where adding resources to a network can lead to a deterioration of performance when selfish, shortest-path routing is employed by individual user flows. This paradox—"how can things be worse when resources are more plentiful?"—cannot arise when routing actions are cooperative. Although the causes underlying Braess's paradox are well understood, the extent to which the paradox manifests itself in Internet routing is unknown. Selfish routing can have a ping-pong-like effect on instability. Users routing traffic over a congested link, upon discovering a newly uncongested link, may switch over in tandem in an attempt to improve individual performance causing the uncongested link to be swamped. This results in a state where the previously congested link becomes uncongested, prompting a reverse migration and consequent oscillatory behavior. In section 5 we will discuss noncooperative network games where players share multiple classes of bounded resources—the congestion control game being a special case where there is a single shared resource—and consider the game structure of noncooperative multi-class QoS provisioning.

Organizational Behavior. An important variable of Internet traffic engineering is the organizational behavior of autonomous systems which intimately impacts routing, QoS, topology, and deployment of new traffic management solutions. Consider the packet forwarding process at IP routers in the packet journey example. When a user downloads a file from a Web server, packets carrying the content of the file are routed by two separate subsystems: interdomain routing and intradomain routing. Supposing the client and server machines reside on different domains—which is generally the case, although caching tries to place frequently accessed content close to where the demand is—the path undertaken by packets at the granularity of domains is determined by Border Gateway Protocol (BGP), an interdomain routing protocol that governs packet forwarding across ASes. The path chosen by a packet as it traverses IP routers within a domain is determined by intradomain routing protocols. Two examples are Open Shortest Path First (OSPF) and Routing Information Protocol (RIP). OSPF implements Dijkstra's algorithm [42] for computing shortest paths—a centralized method—and RIP implements Bellman-Ford's algorithm [13, 55], which is decentralized. Whereas intradomain routing follows the shortest-path principle, interdomain routing is policy based, meaning that an organization can inject economic, political, and other criteria it deems relevant, including shortest-path, when making routing decisions. This can lead to scenarios where a company, when sending an e-mail to another company across the street, has its message routed to a different continent before it eventually reaches its destination just a holler away. Policy influences peering relations between ASes—who is connected to whom—the substrate upon which interdomain routing operates. Organizational behavior implicitly gives rise to power-law connectivity of Internet AS topology.

Organizational behavior affects the viability of new Internet technology deployment, especially as it pertains to quality of service. The service quality experienced by a user is end-to-end, encompassing end systems and their resources— that is, Central Processing Unit (CPU) speed and access bandwidth—and the state of intermediate hops that packets must traverse to reach their destination. End-to-end QoS is only as good as the weakest link in the resource chain whose components may span multiple autonomous systems under the governance of different organizations. To achieve guaranteed QoS, say, in the form of a dedicated 128 Kbps (kilo-bit-per-second) communication channel for CD quality real-time audio, all ASes on an end-to-end path must participate and reserve the required resources. Although the technology for performing the necessary coordination and signaling is there, such services presently do not exist on the Internet outside of limited settings such as leasing of lines from a single service provider and bandwidth commodity markets that trade raw bandwidth without the nuts-and-bolts needed to achieve on-demand QoS across domains. In telephony, an international call will traverse multiple carriers, but agreements exist that allow a dedicated end-to-end channel to be set up on the fly. In the airline industry, codesharing allows multiple carriers to route passengers across their collective routes, reserving a seat on each leg of an end-to-end journey. Of course, we are ignoring overbooking, flight delays, and other factors that contribute to the end-to-end flying experience. On the Internet, policy barriers between administrative domains prevent QoS solutions from being realized, prompting a revival of application layer methods that package services that rely only on IP's reachability functionality. Giving up on the ability to exercise direct resource control facilitates deployability, however, at the cost of performance.

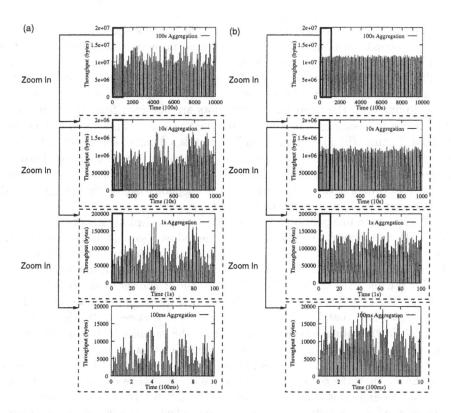

FIGURE 1 Self-similar burstiness. (a) Long-range dependent network traffic. (b) Short-range dependent Poisson traffic.

3 SELF-SIMILAR TRAFFIC

3.1 WHAT IS SELF-SIMILAR TRAFFIC?

Suppose we instrument a router so that we can monitor and record the number of bytes (or packets), per unit of time, that exit a link. The time unit, for example, may be 10 milliseconds, which implies that every logged measurement represents the total number of bytes that have left through the link during a 10-millisecond interval. This results in a time series $X(t)$ with t discrete. Let $X^{(10)}(t)$ denote the time series arising from the same measurement process with the difference that the time unit is 100 msec, that is, ten-fold more coarse. Similarly, $X^{(100)}(t)$ denotes measurement at the granularity of 1 sec and, in general, we define the aggregated time series at aggregation level m as $X^{(m)}(i) = \sum_{t=m(i-1)+1}^{mi} X(t)/m$ which is composed of averaged, non-overlapping m-blocks.

Figure 1(a) shows TCP/IP traffic where the top plot shows measurements at the 100-second granularity over a 10,000-second measurement period. Hence the plot contains 100 data points or samples. The second plot from the top shows traffic logs at the 10-second granularity, where the measurement data are taken from the first 1,000-second interval as indicated by the rectangular time window. In a similar vein, the third and fourth plots are obtained from their preceding time series by zooming in on the first ten data points and magnifying the detail tenfold. Figure 1(b) shows corresponding multiscale plots for Poisson traffic that is more representative of telephone traffic. We observe a stark difference between figure 1(a) and figure 1(b): network traffic variability or burstiness is preserved across four orders of time scale, whereas for telephony-like traffic the aggregated plots become rapidly flat with increasing time scale.

If $X(t)$ were independent, in particular, independent identically distributed (iid) with finite variance σ^2, $X^{(m)}$ can be viewed as a sample mean and its variance is given by $\text{Var}(X^{(m)}(i)) = \sigma^2/m$. Thus, as a function of the sample size, that is, aggregation level m, the rate of decay is polynomial in m with exponent -1. We estimate the shape of $\text{Var}(X^{(m)})$ for the traffic series in figure 1(b) and find that $\text{Var}(X^{(m)}) \propto m^{-\beta}$ with β close to 1. If we do the same for figure 1(a), however, we find that $\text{Var}(X^{(m)}) \propto m^{-\beta}$ where β is closer to 0. Thus, for network traffic exhibiting self-similar burstiness, the variance at larger time scales diminishes, albeit with a rate that is slower than that of independent traffic. This implies that the traffic series in figure 1(a) is correlated—the closer β is to zero the stronger the correlation—indicative of long-range dependence. By convention, the exponent β is denoted as $\beta = 2 - 2H$ where H is called the Hurst parameter after the hydrologist Hurst, who studied reservoir capacity planning using Nile River data [66]. H close to $1/2$ is associated with short-range correlation, and H close to 1 is indicative of long-range correlation. If we estimate the autocorrelation function $r^{(m)}(k)$ for the time series in figure 1(a) where k is the time lag, we find that

$$r^{(m)}(k) \approx r(k) \propto k^{2H-2} \tag{1}$$

with H close to 1. It is in the sense of (1)—the correlation structure across time scales is preserved—that long-range dependent traffic is self-similar. In mathematical modeling of self-similar traffic, we may consider second-order stationary $X(t)$ whose autocovariance satisfies

$$\gamma^{(m)}(k) = \frac{\sigma^2}{2}\left((k+1)^{2H} - 2k^{2H} + (k-1)^{2H}\right) \tag{2}$$

for all k and m. $X(t)$ is called exactly second-order self-similar with Hurst parameter $1/2 < H < 1$. Equation (2) implies (1) where the autocorrelation structure is exactly preserved. When (2) holds asymptotically in m, $X(t)$ is called asymptotically second-order self-similar. Property (1) also holds only asymptotically, that is, $r^{(m)}(k) \sim r(k)$. We refer the reader to [17, 107] for a more comprehensive discussion.

3.2　WHAT CAUSES TRAFFIC SELF-SIMILARITY?

3.2.1　Structural Traffic Models.

If we compare the 10-second aggregation plots of figure 1(a) and figure 1(b), we find that their average is about the same. However, the left plot possesses significant undulations whereas the right plot is fairly even. In the left plot there are 10-second intervals where traffic demand significantly exceeds the average and time intervals where demand significantly falls below. Matching supply, that is, resources, to demand requires correspondingly variable resources that can be adjusted to the time-varying demand if loss or waiting during peak periods is to be avoided. Frequent reshuffling of resources can be difficult and, in some cases, such as installing physical lines and routing equipment, outright infeasible. If resources are allocated at the peak rate then customer satisfaction is assured, but at the cost of wasted or underutilized resources during lull periods, which may translate to increased resource/service prices. Utilization can be improved by lowering the allocated bandwidth, albeit at the expense of quality of service. Bandwidth and buffer space—the two principal network resources—are not storable commodities. That is, when a 100 Mbps link is utilized 50% during an hour, 50 Mbps of unused bandwidth is not saved and available for consumption during the next hour for a total bandwidth budget of 150 Mbps. The use-it-or-lose-it nature of network resources—similarities exist with perishable goods—complicates resource provisioning when traffic demand is bursty across a wide range of time scales. In the case of flat traffic, it is possible to "have the cake and eat it too": both high QoS and high utilization are attainable.

Traffic self-similarity is not a mere curiosity but has potentially serious ramifications for network engineering. Thus, understanding its causes is important. Let us consider an abstraction of a user's traffic flow, called the on-off model, in which a user alternates between on (active) and off (inactive) states. During an on-period, data transmission is assumed underway—by default, at a constant rate—which is also referred to as a packet train [70]. During an off-period, the session is assumed to be idle with no traffic emitted. When viewed as a stochastic process, the on-off model can be described by a set of random variables $\tau_{on}(1), \tau_{off}(1), \tau_{on}(2), \tau_{off}(2), \ldots$ denoting the lengths of the first on-period, the first off-period, the second on-period, and so forth. The on-period random variables are assumed to be iid and independent of the off-period random variables, and similarly for off-period random variables. An instance of an on-off traffic process is defined by the distributions of τ_{on} and τ_{off}. The collective traffic impinging at a router's bottleneck link, which is comprised of several user flows, may be modeled as a superposition of multiple on-off processes. This is depicted in figure 2.

Traffic measurements have shown that on-periods tend to be heavy-tailed while off-periods are light-tailed [142]. A probability distribution is heavy-tailed

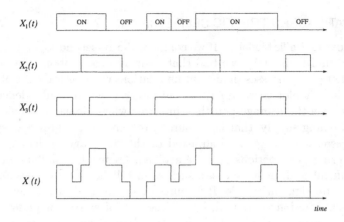

FIGURE 2   Superposition of three on-off processes.

with index $0 < \alpha < 2$ if the tail of the distribution follows a power law

$$\Pr\{Z > x\} \sim c x^{-\alpha} \tag{3}$$

for large x. The precise mathematical counterpart of heavy-tailed random variables are regularly varying random variables [90] whose definition involves the use of slowly varying functions, a technical detail we will ignore here. A canonical heavy-tailed distribution is the Pareto distribution whose distribution function is given by $\Pr\{Z \leq x\} = 1 - (b/x)^{\alpha}$, $b \leq x$, where $0 < \alpha < 2$ is the shape parameter (or tail index) and b is called the location parameter. The Pareto distribution has a power-law tail for all $x \geq b$. Heavy-tailed random variables possess infinite variance, and if $0 < \alpha \leq 1$, they also have an unbounded mean. For traffic modeling purposes, we are interested in the regime $1 < \alpha < 2$. Practically, an on-off process with heavy-tailed on-periods and light-tailed—by default, exponential—off-periods gives rise to many short traffic bursts mixed in with a few very long transmissions. This captures the empirical fact that most TCP sessions are short-lived whereas a few are long-lived. Superposition of independent on-off processes with heavy-tailed on periods leads to asymptotic second-order self-similarity [131], in particular, fractional Gaussian noise [87], a generalization of Gaussian noise ($H = 1/2$). An equivalence relationship holds where the superposition of on-off processes is long-range dependent ($H > 1/2$) if and only if the on-periods—or the off-periods—are heavy-tailed ($1 < \alpha < 2$). For this reason, in the second-order self-similarity context, long-range dependence—technically defined as having a non-summable autocorrelation function—and self-similarity are used interchangeably.

An intimately related, perhaps even more succinct traffic model leading to second-order self-similarity is the $M/G/\infty$ traffic model. We think of traffic as

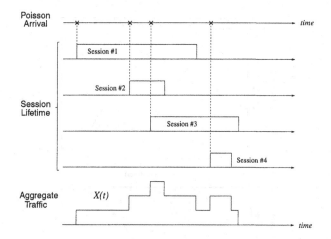

FIGURE 3 $M/G/\infty$ traffic model.

being generated by Poisson arrivals—the inter-arrival time between successive sessions is exponentially distributed—where the lifetime of sessions is heavy-tailed.

A session is viewed as a single on-period. This is illustrated in figure 3. Customers arrive randomly, and most customers are smallish ("mice") whereas a few are very big ("elephants"). $M/G/\infty$ denotes a queuing system, so its interpretation as a traffic model needs some explanation [35, 110]. $M/G/\infty$ describes a queuing system where customers arrive randomly—in other words, their inter-arrival time is exponentially distributed (the "M" in Kendall's notation which standards for Markovian)—there is an infinite number of servers (the "∞"), and a newly arriving customer is assigned to one of the idle servers. At any given time only a finite number of servers are busy. The time to service a customer can follow any distribution (the "G" which stands for general). We are specifically interested in heavy-tailed service times. The performance variable of interest in $M/G/\infty$ is the counting process $X(t)$ which at time t tallies how many servers are busy. It is readily checked that the busy server process $X(t)$ of an $M/G/\infty$ queue with heavy-tailed service times corresponds to the traffic model where session arrivals are Poisson and session lifetimes are heavy-tailed. Again, the essential ingredient that renders the $M/G/\infty$ traffic model long-range dependent is the heavy-tailedness in the service time.

3.2.2 Causality: Heavy-tailed File Sizes.
Since empirical measurement data pointed to heavy-tailed session lifetimes [110, 142], a question remained as to why this is the case. This was addressed in Park et al. [108] where UNIX file systems research carried out in the 1980s was examined and shown to indicate

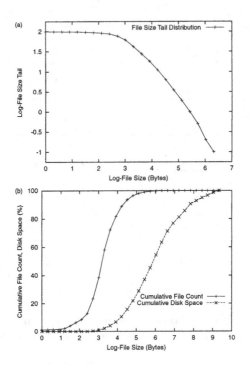

FIGURE 4   Heavy-tailed UNIX file size distribution. (a) Log-log plot of file size tail. (b) Cumulative distribution of file size count and disk space.

skews in file sizes consistent with heavy-tailedness. In Cunha et al. [39], file access by Web clients at Boston University during Nov. 1994–Feb. 1995 was shown to be heavy-tailed. The role of UNIX file systems for explicating the causality of self-similar network traffic is relevant for two reasons. First, the original Bellcore data, the basis for Leland et al.'s seminal study [81], was collected during 1989–1992 when the World Wide Web did not exist yet: the Web could be a facilitator but not the root cause of traffic self-similarity. Second, empirical evidence from the 1980s in the file system research community predating the self-similar traffic discovery provided independent support that heavy-tailedness of file sizes may be a phenomenon that is not necessarily specific to the Internet.

Figure 4(a) shows the log-log tail distribution of file sizes from a 1993 survey of UNIX file systems [68]. For large file sizes—the detailed structure of small files has negligible influence on long-range dependence—we find a straight-line fit consistent with heavy-tailedness. Figure 4(b) shows the cumulative distribution of the percentage of files of a certain size and the corresponding cumulative disk space consumed. File size is shown in log-scale with base 10. We observe that close to 90% of files are of size less than 10,000 bytes. Collectively, they consume less than 10% of the disk space. The minority of files that are bigger than 10,000 bytes—some are very large—consume the bulk of the disk space. Together they invoke the metaphor "many mice and a few elephants." We will encounter power-law skews again when discussing Internet connectivity in section 4. The relevance of Cunha et al.'s Web access measurements [39] (and their further analysis in Crovella and Bestavros [36]) derives from the fact that file access is not synonymous with file size distribution: a file server may be heavy-tailed, but if users only access small files, then the file access distribution—not file size distribution—becomes the variable of interest. The results in Cunha et al. [39] showed that user access behavior resembled random sampling from a heavy-tailed file size distribution.

3.2.3 Influence of Traffic Control. To complete the reductionist reasoning and causal chain—(i) traffic is self-similar because multiplexing of sessions with heavy-tailed lifetimes leads to self-similarity, (ii) session lifetimes are heavy-tailed because most traffic is TCP file transfer traffic and file sizes are heavy-tailed, ergo (iii) traffic is self-similar because file sizes are heavy-tailed—required showing that TCP does not significantly interfere with the transfer process of heavy-tailedness of file size distribution into heavy-tailedness of session lifetime. In Park et al. [108] it was shown that TCP approximately preserved this transfer relationship, and coupling of TCP sessions at shared bottleneck links did not break the chain. On the other hand, User Datagram Protocol (UDP) traffic that is not feedback controlled—UDP is a protocol operating at the same layer as TCP, the essential function of which is to identify the application layer process that a packet is destined to—resulted in reduced self-similarity during contention periods. This can be understood by noting that TCP conserves information due to its retransmission based reliability mechanism, whereas UDP—unless application layer protocols running above it implement their own reliability mechanism—suffers information loss which diminishes the impact of the file size tail. During high contention accompanied by persistent packet loss, TCP stretches its transmission into an on-average thin stream where the reduced signal strength stemming from thinning is compensated by the lengthened file transfer completion time. The presence of duplicate retransmissions may amplify the total number of bytes transmitted by TCP. In the case of plain UDP, this "conservation law" need not hold.

An on-average flattish TCP session contains internal structure at the time scale of the round-trip time (RTT), in other words, network latency, at which feedback control actions are undertaken. The output behavior of a typical long-lived TCP session follows the shape of a sawtooth interspersed with varying lull periods. Linear ascent in the sawtooth is affected by TCP's feedback control which additively increases traffic submitted to the network when unused bandwidth is deemed available. Sharp descent in the sawtooth pattern stems from TCP's multiplicative back-off, which is instituted at times of perceived packet

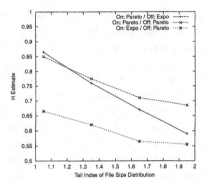

FIGURE 5 Hurst parameter of on/off TCP sessions: Pareto file size/exponential idle period, Pareto file size/Pareto idle period, and exponential file size/Pareto idle period.

loss. The lull periods are introduced by TCP's retransmission mechanism, which injects exponentially increasing pause periods during successive retransmission attempts. Small time-scale TCP dynamics can lead to chaotic trajectories and contribute to longer time-scale correlation [136, 137]. In particular, both exponentially increasing idle periods—extended inactivity is a form of correlation and creates sender-side backlog—and linear growth of throughput that elongates transmission duration contribute to long-range correlation. The magnitude of this contribution, however, is small when compared to the impact of heavy-tailedness in inducing long-range dependence of network traffic. This can be seen in figure 5 (based on results from Park et al. [108]) in which the Hurst parameter of aggregate traffic stemming from 32 concurrent TCP connections that share a bottleneck link is shown for three different cases: (i) file sizes are drawn from a Pareto distribution with the specified tail index and idle periods between successive file transfer sessions within a connection are set exponential; (ii) both file size distribution and session inter-arrival times are set to be Pareto; and (iii) file sizes are set to be exponential but idle times are set to be Pareto. Case (i) corresponds to the canonical configuration of practical interest reflecting Internet workload measurements. In case (ii), Pareto inter-session arrival times within a TCP connection inject an additional measure of long-range dependence that for lighter tails (e.g., $\alpha = 1.65$ and 1.95) exact a noticeable effect. For heavier tails (e.g., $\alpha = 1.05$ and 1.35), however, the effect is negligible. The most relevant scenario is case (iii) where exponential file sizes remove the heavy-tailed file size factor of long-range dependence, while Pareto idle times inject a long-range correlation stronger than those induced by TCP's exponential backoff. As the tail index approaches 1, we observe an increase in the Hurst parameter. However, its magnitude (around 0.65) is significantly smaller than that of case (i) (around 0.85) representative of Internet traffic. It is for this reason that heavy-tailed file size distribution constitutes the dominant cause of long-range dependence in Internet traffic, with TCP dynamics playing a tertiary role.

3.3 PERFORMANCE IMPLICATIONS OF SELF-SIMILAR BURSTINESS

3.3.1 Queuing: How Long Must One Wait.

A key concern of network performance evaluation is how long packets arriving at routers have to wait before they are processed and sent on their way. The same is true of client requests at servers. From everyday experience at toll booths, restaurants, and fast food pick-ups, we know that the waiting time is influenced by the bursty nature of arrivals— sometimes attributed to "luck" or the lack thereof—in addition to time-of-day and overall traffic intensity. Such phenomena are studied in queuing theory [76], where the biggest challenge lies in understanding the sometimes complicated structure of bursty arrivals and their impact on waiting. To understand the influence of heavy-tailedness in the input on queuing, we will consider a simplified queuing system with traffic input $X(t)$, storage occupancy $Q(t)$, service rate μ, and storage capacity S, where packets are treated as fluid and $S = \infty$. The motion of the system is governed by

$$Q(t + \Delta t) = \max\{Q(t) + X(t, t + \Delta t) - \mu \Delta t, \, 0\} \tag{4}$$

where $X(t, t + \Delta t)$ denotes the input during the specified time interval and the queue drains at constant rate μ. Thus, storage occupancy at Δt in the future is determined by adding the net influx, $X(t, t + \Delta t) - \mu \Delta t$, to the current occupancy level. Equation (4) defines a random walk with a reflecting barrier at zero. To prevent the queue from growing out of bounds, we require that there be a negative drift, $\lambda < \mu$, where λ is the average traffic rate of input $X(t)$. Assuming the system is well behaved, we seek to find the equilibrium distribution of $Q(t)$, $Q(\infty)$, the main performance variable. When the distribution of $Q(\infty)$ has an exponential tail, in other words, $\Pr\{Q(\infty) > x\} \propto e^{-ax}$, buffer dimensioning becomes cost-effective since an additional unit of buffer capacity multiplicatively decreases the probability of overcrowding. The bulk of Markovian or short-range dependent traffic leads to $Q(\infty)$ with light tails. For long-range dependent (LRD) traffic, the picture is starkly different: heavy-tailedness in the input is preserved and translated to heavy-tailedness in the queue length distribution. Thus the probability of overcrowding for large x is significantly amplified, rendering buffer dimensioning a less cost-effective resource provisioning strategy for accommodating crowds. We will illustrate this transfer relation using a single on-off process with heavy-tailed on-periods and light-tailed off-periods.

Let the on-period be Pareto with tail index α and location parameter b, and let the off-period be exponential with parameter λ_{off}. Let λ_{on} denote the traffic rate during an on-period. Then $\lambda = \lambda_{\text{on}} E\{\tau_{\text{on}}\}/(E\{\tau_{\text{on}}\} + E\{\tau_{\text{off}}\})$, and $\lambda < \mu < \lambda_{\text{on}}$ is the regime where queuing behavior is nondegenerate: neither unbounded nor empty. Suppose we want to lower bound the tail probability $\Pr\{Q(\infty) > x\}$ for large x. Let $\tau_{\text{on}}(1), \tau_{\text{off}}(1), \tau_{\text{on}}(2), \tau_{\text{off}}(2), \ldots, \tau_{\text{on}}(n), \tau_{\text{off}}(n)$ denote a block

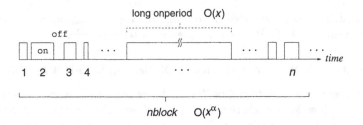

FIGURE 6   Block of n alternating on- and off-periods at time scale $n = O(x^\alpha)$.

of n alternating on- and off-periods where $n = y^\alpha/b^\alpha$ and $y = x/(\lambda_{\text{on}} - \mu)$. Since $\Pr\{\tau_{\text{on}} > y\} = (y/b)^{-\alpha}$, the expected number of on-periods in the n-block that exceed y is one. Figure 6 depicts the n-block configuration. Let L denote the length of the "long" on-period. Its expectation is given by

$$E\{L\} = E\{\tau_{\text{on}} \mid \tau_{\text{on}} > y\} = \int_0^\infty (s+y)\frac{\alpha y^\alpha}{(s+y)^{\alpha+1}}ds = \frac{\alpha}{\alpha-1}y$$

where $\alpha y^\alpha/(s+y)^{\alpha+1}$ is the conditional probability. y is set to $x/(\lambda_{\text{on}} - \mu)$ so that the queue is assured to build up to x. Since $\lambda_{\text{on}} > \mu$ the queue remains above x for at least $E\{L\} - x/(\lambda_{\text{on}} - \mu)$ during the long on-period. Let B denote the length of the n-block. We have

$$E\{B\} = n\left(\frac{1}{\lambda_{\text{off}}} + \frac{k\alpha}{\alpha-1}\right) = \Theta(x^\alpha)$$

where $E\{\tau_{\text{on}}\} = k\alpha/(\alpha-1)$. For large x, we estimate a lower bound using

$$\Pr\{Q(\infty) > x\} \gtrsim \frac{E\{L\} - y}{E\{B\}} = \Theta(x^{1-\alpha}).$$

The queue tail decays polynomially and heavy-tailedness is preserved. The preceding is not a proof but an intuitive illustration using elementary arguments to show why the queue would find itself overcrowded for $\Theta(x^{1-\alpha})$ fraction during a $\Theta(x^\alpha)$-length period. Queuing with long-range dependent input is discussed in Boxma and Cohen [19], Liu et al. [83], and Makowski and Parulekar [85].

An upper bound of $\Pr\{Q(\infty) > x\}$ may be gleaned from the following closure property satisfied by heavy-tailed random variables: for large x,

$$\Pr\{\tau_{\text{on}}(1) + \cdots + \tau_{\text{on}}(n) > x\} \sim n\Pr\{\tau_{\text{on}} > x\}. \tag{5}$$

The most likely time scale at which n heavy-tailed on-periods in a n-block would collude to cause the queue to fill beyond x is $n \sim cx^\alpha$. Moreover, (5) implies $\Pr\{\tau_{\text{on}}(1) + \cdots + \tau_{\text{on}}(n) > x\} \sim \Pr\{\max\{\tau_{\text{on}}(1), \ldots, \tau_{\text{on}}(n)\} > x\}$ which indicates

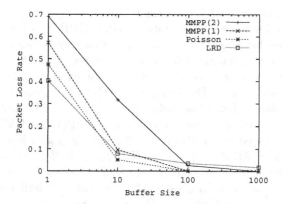

FIGURE 7   Packet loss rate of finite buffer queue with Poisson, LRD, and MMPP input.

that overcrowding occurs "suddenly": at time scale $O(x^\alpha)$ the typical picture is that of a single, long on-period dominating the queue dynamics with respect to storage level x. Hence, $\Pr\{Q(\infty) > x\}$ is upper bounded by a function that is $O(x^{1-\alpha})$ for large x. In the case of exponential on-period $\Pr\{\tau_{\mathrm{on}} \leq x\} = 1 - e^{\kappa x}$, an analogous lower bound argument with $n = e^{\kappa x}$ yields a $\Theta(e^{-\kappa x})$ bound since the conditional expectation of the exponential distribution is constant. For the upper bound, $\Pr\{\tau_{\mathrm{on}}(1) + \cdots + \tau_{\mathrm{on}}(n) > x\}$ is exponentially small with a corresponding exponential upper bound on the queue tail.

3.3.2   Finite Time-scale Effects.   The arguments used for bounding $\Pr\{Q(\infty) > x\}$ required that x be large, that is, $x \to \infty$. Queuing analyses with LRD input that provide rigorous estimates on equilibrium tail probability are asymptotic in nature, a handicap when it comes to resource dimensioning and computing buffer overflow for practical systems with small x. Imposing finite storage capacity, $S < \infty$, complicates matters—the resulting queue process has two reflecting barriers—and drawing definitive conclusions on queuing delay and packet loss rate in finitary systems is difficult. Long-range dependent traffic, when compared with short-range dependent traffic, may or may not affect a higher packet loss rate: it depends on the specifics of the finitary storage system. That a heavy-tailed queue tail is not synonymous with higher packet loss can be seen from figure 7, which shows packet loss rate as a function of buffer size for a simulated queuing system with three types of traffic input: Poisson, LRD, and MMPP. A two-state Markov Modulated Poisson Process (MMPP) is a stochastic process generated by a two-state Markov chain—when the system finds itself in the first state, it behaves as a Poisson process with rate λ_1; when the system is in the second state, it produces Poisson traffic with rate λ_2. By appropriately tuning the transition probabilities along with λ_1 and λ_2, varying degrees of short-range dependent burstiness can be introduced. MMPP(2) denotes MMPP traffic whose parameters have been set to make it significantly burstier than MMPP(1). LRD denotes an aggregated on-off process (see sec. 3.2.1, fig. 2) composed of 32 independent on-off sources each with tail index $\alpha = 1.1$. All four traffic processes—Poisson, LRD, MMPP(1), and MMPP(2)—possess the same traffic rate. Figure 7 shows that when buffer capacity is 1, LRD traffic achieves the lowest packet loss rate, followed by Poisson, MMPP(1), and MMPP(2). When buffer size is increased to 10, Poisson traffic has the lowest loss rate with LRD second, followed by MMPP(1) and MMPP(2). As buffer capacity increases to 100, another inversion takes place where LRD overtakes both MMPP(1) and MMPP(2), resulting in the highest packet loss rate. At buffer size 1,000, only LRD traffic suffers nonnegligible loss. Thus we observe a transition in the packet loss order

$$\text{LRD} \preceq \text{Poisson} \preceq \text{MMPP} \longmapsto \text{Poisson} \preceq \text{MMPP} \preceq \text{LRD}$$

as buffer size is increased from small to large. The reason for this is: when buffer capacity is small, variability and correlation nascent in short-range dependent traffic are sufficient for dominating queue dynamics with respect to buffer overflow; only when buffer capacity is large does long-range dependence have an advantage at overflowing the queue and exerting a deciding influence on packet loss rate vis-à-vis short-range dependence. In the latter, the collusion required for causing overcrowding is exponentially rare and all too brief when it happens.

3.3.3   Self-similar Burstiness and Jitter.

The queuing discussion showed that long-range dependence in self-similar traffic does not necessarily lead to amplified packet loss rate. Although packet loss rate, a first-order performance measure, is a primary yardstick in Internet performance evaluation, another important criterion is "jitter"—a generic term for second-order performance statistics—that captures variability in the packet loss or packet delay process. In multimedia communication, it is not only needed that the average packet loss be small but that the loss pattern be dispersed so that their effect may be masked by the human perceptual system or through traffic control. Masking by traffic control can be effected through forward error correction (FEC), that is, channel coding, where redundant information is transmitted to offset damage that may occur during travel. At the granularity of documents or video frames where a single object is split into several packets, encoder and decoder functions exist [113, 115] that satisfy the k-out-of-N property: k information packets are transformed into $N = k + h$ encoded packets—h represents the degree of redundancy—and a receiver is able to reconstruct the original content as long as no more than h packets, whichever they may be, are lost in the network. FEC is especially relevant in real-time applications over long distances where retransmission based error correction may not be an option. For an FEC-protected traffic stream,

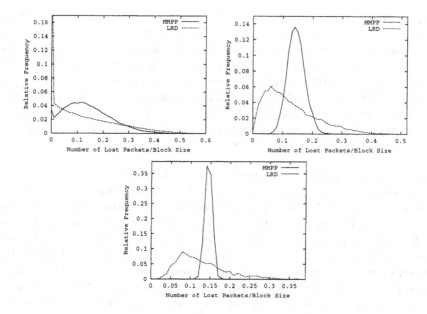

FIGURE 8 Second-order performance: Block loss probability comparison between LRD and MMPP traffic. (a) Block size $N = 100$. (b) $N = 1,000$. (c) $N = 10,000$.

successful recovery is impeded by jitter in the form of bursty or clustered packet loss as this can lead to too many losses within a block.

Given traffic $X(t)$ and block size N, the block loss process induced by feeding $X(t)$ into a finite buffer queue is a sequence of random variables $L_N(i)$, $i \in \mathbb{Z}_+$, where $L_N(i) \in \{0, 1, \ldots, N\}$ denotes the number of packet losses suffered in the ith block. $L_N(\infty)/N \in [0, 1]$ denotes normalized block loss $L_N(i)$ in steady-state as $i \to \infty$. Figure 8 shows normalized block loss distribution when LRD and MMPP traffic are fed into a queue of buffer size 10. The packet loss rate experienced by MMPP is higher than that of LRD, and we interprete buffer overflow with respect to three different block sizes $N = 100, 1,000, 10,000$. Figure 8(b) shows normalized block loss distribution for $N = 1,000$. Even though MMPP's loss rate is higher, LRD's block loss distribution has a wider spread and variance, incurring up to 40% packet loss within some blocks. MMPP's block loss, on the other hand, is bounded by 23%. Assuming the input is already FEC-encoded with redundancy h, $\Pr\{L_N(\infty)/N > h/N\}$ is the probability that the k-out-of-N property is violated and resultant decoding unsuccessful. Thus, LRD's heavier block loss tail, $\Pr\{L_N(\infty)/N > x\}$, $x \in [0, 1]$, implies that FEC performance significantly degrades for LRD traffic when compared to MMPP traffic despite the former's smaller loss rate. The impact of self-similar burstiness on second-order performance is distinct from that of first-order performance. It is affected by block size N, which exerts a similar influence as buffer capacity on block loss performance. Figures 8(a) and 8(c) show that the relative difference in block loss variance between LRD and MMPP becomes pronounced for $N = 10,000$, whereas for $N = 100$ the difference becomes dampened. For smaller N, the block loss distributions of LRD and MMPP grow close with their tails eventually stretching to 100%. For $N = 1$, $\Pr\{L_1(\infty) = 1\} = 1 - \Pr\{L_1(\infty) = 0\}$ is the loss rate. At the opposite extreme, for large N the distribution of $L_N(\infty)/N$ becomes concentrated around the loss rate.

3.3.4 Sampling and Slow Convergence.

Given the instrumental role played by heavy-tailedness in self-similar network traffic, sampling from heavy-tailed distributions is relevant for network simulation including artificial workload generation. A canonical example is the comparison of queuing performance between short-range dependent and long-range dependent traffic, where the latter may involve on-off or $M/G/\infty$ traffic. A key requirement for comparability is that the average traffic intensity of the input be the same so that observed differences in loss performance can be attributed to the correlation structure of the input, not sampling induced discrepancy and bias in the actual traffic rate. Figure 9(a) shows the running sample mean for 60×10^6 samples drawn from an exponential distribution with rate $1/60$ for three different random seeds. Figure 9(b) shows corresponding sample means for a Pareto distribution with parameters $\alpha = 1.2$ and $b = 10$. Both distributions have the same population mean 60, but their convergence properties are markedly different. Whereas the sample mean of the exponential distribution converges rapidly and hugs the population mean after 6,000 samples, the sample mean of the Pareto distribution requires at least 30 million samples to approximate the population mean with 5% accuracy. The polynomial tail amplifies the contribution of large values to the expectation, and its realization requires polynomially many observations with respect to the inverse of the accuracy parameter but greater than exponentially many as a function of the tail index.

To see this, let Z be Pareto with tail index α and location parameter b. The probability density function of Z is given by $f(x) = \alpha b^\alpha x^{-(1+\alpha)}$. Recall that Z has a finite mean but infinite variance for $1 < \alpha < 2$. We split the expectation, $E\{Z\} = b\alpha/(\alpha - 1)$, into two parts separated by y, $E\{Z\} = \int_b^y xf(x)dx + \int_y^\infty xf(x)dx$, and consider the relative contribution of the tail in achieving an accuracy level ε,

$$\frac{\int_y^\infty xf(x)dx}{E\{Z\}} = \frac{\alpha - 1}{b\alpha} \int_b^y x\alpha b^\alpha x^{-(1+\alpha)} dx = \left(\frac{b}{y}\right)^{\alpha-1} < \varepsilon.$$

We have $y_0 = b(1/\varepsilon)^{1/(\alpha-1)} < y$, and since $\Pr\{Z > y_0\} = \left(\frac{b}{y_0}\right)^\alpha = \varepsilon^{\frac{\alpha}{\alpha-1}}$, we require $N_0 = \varepsilon^{-\frac{\alpha}{\alpha-1}}$ samples to expect a value greater than y_0. The number of samples is polynomial in the inverse of the accuracy parameter, $1/\varepsilon$, but grows

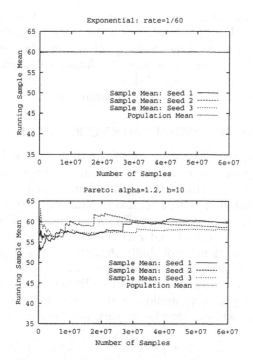

FIGURE 9   Sampling and convergence. (a) Running sample mean for exponential distribution with rate 1/60. (b) Running sample mean for Pareto distribution with $\alpha = 1.2$ and $b = 10$.

faster than exponentially in α as it approaches 1. For $\alpha = 1.2$ and $\varepsilon = 0.05$, N_0 yields 64 million samples. In the case of the exponential distribution, the contribution of the tail in the expectation is exponentially small in y, which implies that y need only be logarithmically large in $1/\varepsilon$. However, due to the exponential smallness of the underlying probability distribution, the number of samples required remains a polynomial function of $1/\varepsilon$, albeit with a small exponent that does not depend on the rate parameter.

Returning to figure 9(b), we observe sudden jumps in the sample mean at 8 million, 17 million, and 26 million samples, which illustrates that the sum of heavy-tailed random variables is dominated by the maximum "outlier," a consequence of property (5) (see sec. 3.3.1). The slow convergence associated with heavy-tailed random variables exacts a heavy toll on sampling, and speed-up methods that admit shortcuts—the subject of rare event simulation [63]— are needed. For example, importance sampling, a technique that modifies an underlying probability measure to pump up the likelihood of rare events, is aimed at achieving both small variance and relative error. Importance sampling with an exponential change of measure has been applied, with some success, to light-tailed distributions, but encounters difficulties when extending to heavy-tailed distributions. On the other hand, a change of measure that selects distributions with heavier tails shows initial promise [6]. Additional discussion on sampling and fast simulation with heavy-tailed and long-range dependent input can be found in Crovella and Lipsky [37] and Gallardo et al. [57].

3.4 TRAFFIC CONTROL: WHAT CAN BE DONE ABOUT IT?

3.4.1 Large Time-scale Predictability.
Given that long-range dependence in self-similar traffic is caused by heavy-tailed session durations which, in turn, arise from heavy-tailed file sizes, the predictability inherent in heavy-tailed distributions—manifested as long-term correlations in network traffic—may be harnessed for traffic control purposes. To see why heavy-tailedness implies predictability, consider a heavy-tailed random variable Z with index α. We are interested in the conditional probability $\Pr\{Z > x + y \mid Z > y\}$, which captures the likelihood of predicting the "future" value of Z given its "past" value y. For example, if Z were the session duration of an IP flow that is being tracked at a router, $\Pr\{Z > x + y \mid Z > y\}$ would give the probability that a session that has lasted for y seconds will continue for at least another x seconds. We have

$$\Pr\{Z > x + y \mid Z > y\} = \frac{\Pr\{Z > x + y\}}{\Pr\{Z > y\}} \sim \left(\frac{y}{y+x}\right)^{\alpha}. \tag{6}$$

$\Pr\{Z > x + y \mid Z > y\} \to 1$ as $y \to \infty$, which implies that conditioning the future on an ever longer past brings about certitude. In contrast, for the memoryless exponential distribution, conditioning on the past has no bearing on the future. The expected future duration is given by the conditional expectation $E\{Z \mid Z > y\} = y\alpha/(\alpha - 1)$ (see sec. 3.3.1). Thus, persistence into the future is proportional to persistence in the past, with α close to 1 amplifying the predictable time horizon. The conditional variance, $\mathrm{Var}\{Z \mid Z > y\}$, is unbounded for $0 < \alpha < 2$. In the case of aggregate traffic, for example $M/G/\infty$ input, aggregate traffic at time scale $O(t^{\alpha})$ is dominated by a single long session of duration $O(t)$. Relative signal strength, $t^{1-\alpha}$, decays slowly which renders long-term positive correlation generated by heavy-tailedness detectable in the presence of short-term fluctuations.

Figure 10 shows predictability in aggregate network traffic generated by multiple TCP connections that share a common bottleneck link. We consider two scenarios: (1) files transported by TCP are Pareto with $\alpha = 1.05$, and (2) file size distribution is Pareto with $\alpha = 1.95$. A time series representing traffic measurement is partitioned into two-second time windows. The traffic volume within a time window is quantized into eight levels where one is low and eight is high. Let S_L and S_R denote random variables representing the quantized traffic volume at two consecutive time windows. Figure 10(a) shows the estimated conditional probability $\Pr\{S_R \mid S_L = h\}$ as a function of traffic level

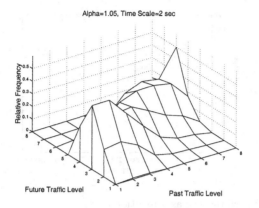

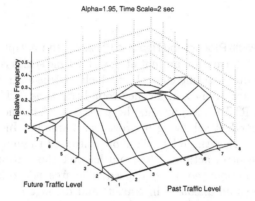

FIGURE 10   Long-range correlation and predictability. (a) Conditional probability for $\alpha = 1.05$ traffic. (b) Conditional probability for $\alpha = 1.95$ traffic.

$h \in \{1, 2, \ldots, 8\}$ for TCP traffic with $\alpha = 1.05$. At each traffic level, we observe a skewed distribution: when S_L is high (low), it is likely that S_R is high (low). Positive correlation between S_L and S_R leads to a diagonally shifting distribution peak in the three-dimensional plot. Figure 10(b) shows the corresponding results for $\alpha = 1.95$. We observe a stark difference: the shape of $\Pr\{S_R \mid S_L = h\}$ does not depend on S_L, for example, $\Pr\{S_R \mid S_L = h\} \approx \Pr\{S_R\}$ for all h. Short-range dependent traffic possesses a conditional probability profile similar to figure 10(b) at the two-second time scale. At time scales one to two orders of magnitude smaller, however, short-range dependent traffic exhibits conditional probability plots similar to that of figure 10(a). That is, at sufficiently small time scales, short-range dependent traffic is predictable.

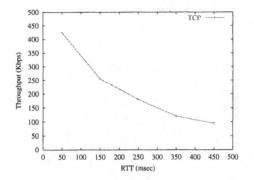

FIGURE 11   TCP performance as a function of RTT.

3.4.2 Delay-bandwidth Product and Reactive Penalty.

In multiple time-scale traffic control [104, 132], information present at multiple time scales is engaged to affect traffic control. Feedback traffic control, of which TCP is an instance, acts at the time scale of feedback latency, also called round-trip time (RTT). Network state information from RTT time units in the past is used in the present to enact control actions aimed at achieving improved performance. In dynamic network environments, the larger the RTT, the more outdated the information, and the less effective the control action ("the train has left the station"). The reactive cost is especially pronounced in high-speed wide-area networks (WANs) where many packets are simultaneously in transit, and damage ensuing from delayed reaction can be significant. For example, in coast-to-coast transmissions, RTT is in the tens of milliseconds and individual broadband access speed can exceed 1 million-bits-per-second (Mbps). In satellite networks, bandwidth is relatively small but two-way latency reaches 500 msec.

Figure 11 shows the diminishing throughput of TCP as RTT increases. Exposure of feedback traffic control to the performance limitation imposed by large delay-bandwidth product networks is an inherent problem. In open-loop traffic control, resources are reserved in advance to protect a traffic flow from uncertainties of future network states in a first-come-first-served (FCFS) shared network environment. When performance guarantees are required, open-loop control is unavoidable. However, it carries its own cost of having to know the characteristics of the traffic flow—not always easy to do—and potential resource underutilization stemming from per-flow reservation near the peak, as opposed to average, data rate. In multiple time scale traffic control, the goal is to exploit long-range predictability in closed-loop traffic control to mitigate the outdatedness of feedback information at the time scale of RTT.

3.4.3 Workload-sensitive Traffic Control.

Two issues need to be addressed when engaging long-range correlation for traffic control: prediction of future traffic and utilization of this information for traffic control. At time t, the average future traffic level at time horizon $t^* > t$, given by

$$\bar{X}(t,t^*) = \frac{1}{t^* - t} \sum_{s=t}^{t^*} X(s),$$

may be predicted using past observations $X(t-1), X(t-2), \ldots, X(t-t_*)$ extending t_* time units into the past. The prediction error $E\{(\bar{X}(t,t^*) - \hat{X}(t,t^*))^2 \mid X(t-1), \ldots, X(t-t_*)\}$ of the best linear unbiased estimator (BLUE) is known in the asymptotic case (i.e., as $t_* \to \infty$) [16]; the finitary case, however, is more difficult [17]. A simpler, but suboptimal, prediction method uses average past traffic level $\bar{X}(t_*, t-1)$ to predict the future: $E\{\bar{X}(t,t^*) \mid \bar{X}(t_*, t-1)\}$. A quantized variant of the conditional expectation predictor has been employed in Park and Tuan [104] and Tuan and Park [132, 133]. Since the aggregate traffic level at a bottleneck router is not directly observable by the sender, active or passive probing must be used—unless routers are enabled to convey their state information directly, which has its own problems—to estimate contention level. In active probing, probe packets are transmitted to ascertain the network state from the packets' delay and loss characteristics observed at the receiver. In passive probing, the characteristics of transmitted application traffic are used to infer the network state. This eliminates additional messaging overhead, albeit at the cost of reduced accuracy. The passive probing method used in Park and Tuan [104] and Tuan and Park [132] utilizes the output behavior of TCP, observable at the sender, to infer the contention level on a bottleneck path. This method relies on the tracking ability of feedback congestion control, including TCP, the output behavior of which can be shown to be negatively correlated to the contention level.

Assuming the average future traffic level is known, how should this information be utilized to effect improved traffic control? Long-term state information, by definition, is slowly varying—short-term fluctuations may be missed—and the predicted information is probabilistic: with some likelihood the predictions are wrong. Hence, at a minimum, a large time scale control action should not do more harm than good. In proactive congestion control, long-term prediction is used to modulate bandwidth consumption behavior at the time scale of RTT: aggressive when the outlook is good and conservative if the outlook is bad. The net effect is improved throughput, achieved by desensitizing control actions against short-term fluctuations while increasing awareness of persistent state changes. By reducing futile reactions to transient events that, by definition, are fleeting and outdated by the time control actions take effect, throughput degradation may be alleviated.

In the context of TCP congestion control [69], long-term prediction may be utilized in two complementary ways. When network conditions are favorable—

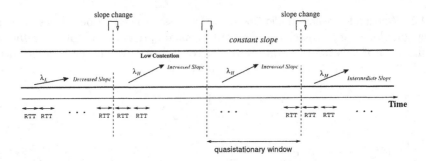

FIGURE 12 Selective slope control: TCP's linear increase slope is modulated as a function of long-term network state.

that is, available bandwidth is high—TCP should aggressively soak up unused bandwidth as the opportunity cost for not doing so is commensurately high. This may be accomplished by increasing the linear rate at which TCP opens up its throttle. Eventually, TCP reaches its maximum throttle or the increased sending rate causes losses at bottleneck routers. In the latter, upon detecting potential packet loss—the TCP sender uses lack of timely acknowledgement from the TCP receiver as an indicator of loss—TCP clamps down on the throttle by a multiplicative factor of 1/2. Consecutive multiplicative clamp-down leads to exponential back-off which, in general, is needed to achieve stability. During periods when available bandwidth is plentiful, however, back-off need not be as drastic. Conversely, when the overall contention level is high, a more conservative bandwidth consumption behavior may be undertaken. A specific form that modulates the slope during linear increase is illustrated in figure 12. Stability of multiple time scale TCP holds as long as TCP's feedback congestion control is stable and the time scale at which long-term control parameter modulation is undertaken significantly exceeds the time scale of RTT. If both conditions are satisfied, time scale separation assures that TCP feedback control reaches equilibrium during a successive pseudo-stationary time window at which control parameters are held constant. In the long run, TCP moves from equilibrium to equilibrium across successive pseudo-stationary time windows.

Figure 13 shows the performance gain achieved by TCP when it is augmented to utilize long-term predicted contention to modulate its linear increase-exponential decrease behavior. We observe that the multiple time scale version of TCP, TCP-MT [104], improves on the throughput of TCP, where throughput gain (%) amplifies with RTT. The quantitative gain depends on the underlying TCP flavor—for example, TCP Reno, NewReno, Vegas—and specific network conditions. Performance gains up to 60% have been observed in prototype systems. Similar results hold for rate-based congestion control which is employed in UDP-based traffic streaming [132]. Feedback congestion control modulation using

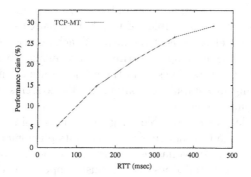

FIGURE 13   Performance gain of multiple time scale TCP: TCP-MT.

long-term information applies to long-lived flows. For short-lived flows, feedback is relevant for error and packet loss recovery. Predictability from heavy-tailedness admits on-line classification of short-lived and long-lived flows so that long-term control actions can affect long-lived sessions in which they matter. The same holds for traffic shaping and admission control: the few elephants that consume much of the Internet bandwidth must be reined in to make a difference. The mice, collectively, exert only a small impact on system performance.

Workload-sensitive traffic control has been applied for real-time video/audio transport using adaptive forward error correction (AFEC), where redundancy is adaptively injected to protect against packet loss without necessitating retransmission [105, 106]. Multiple time-scale AFEC, AFEC-MT, shields AFEC against transient fluctuations, thereby increasing the recovery rate—that is, correct decoding of video/audio frames—at the receiver [133]. In Östring et al. [100], router-aided rate control for self-similar traffic is explored. Sampling based prediction and scheduling is studied in Wu et al. [144]. Heavy-tailedness has been exploited for dynamic load balancing of UNIX processes where process lifetimes are observed to be heavy-tailed [61], enhancing routing stability through long-lived and short-lived IP flow separation where routing updates are desensitized against short-lived flows [124], and process scheduling where preference is given to short-lived processes [38].

3.5   DISCUSSION

The global Internet operates primarily as a distributed client/server system, where the bulk of events entail fetching files of various types—Web page, image, video, and software—from remote sites. Heavy-tailedness of file size distribution is a structural property of distributed systems, an empirical "law" on a par with Poisson session arrival and locality of reference. Memoryless inter-session arrival has enabled Markovian analysis of real-world systems, including capacity planning in telephony, the principles of which date back to Erlang's pioneering work [49]. Locality of reference is at the heart of caching, a resource management technique without which economic information processing would be severely impeded. Collectively, the three laws form the cornerstones of effective system design and engineering in computing/communication systems, with heavy-tailed file size being the junior member.

Heavy-tailed files are responsible for generating self-similar network traffic, an emergent trait of the Internet that transcends its phenomenological curiosity. That is, self-similar burstiness has repercussions to network performance, planning, and control. Heavy-tailedness is a robust property in the sense of being a conserved property: it manifests itself as long-range dependence when channeled across a network, surfaces as heavy-tailed queuing delay when fed into a router's buffer, and translates into long-term predictability when harnessed for traffic control. Heavy-tailed file size is like an invariant that morphs into other heavy-tailed network phenomena but is not readily suppressed.

Fractal properties of network traffic in the form of $1/f$ noise and chaotic dynamics of TCP's nonlinear feedback control are not unrelated (see sec. 3.2.3), but are of secondary import to Internet engineering. The latter provides grounding in real-world considerations that curb chaos-centric interpretations of the Internet when viewed as a complex system. Self-similar traffic is a representative example where a "physics of network traffic" is established spanning measurement, network mechanics, and mathematical modeling. A strength of the Internet is that it comes with concrete application driven issues where complex systems notions exert a tangible, pragmatic influence. These range from structural causes of correlation at a distance, to slow convergence and equilibria, to prediction and control. As an engineered physical system, the Internet affords precise quantitative measurement, allowing testing of theories with respect to cause and effect that may involve complex what-if scenarios, the scope of which goes beyond those feasible in biological, social, and natural physical systems.

The origin of self-similar traffic traces back to heavy-tailedness of file sizes. Venturing a step further, one may ask: why are files heavy-tailed? This is not an uninteresting question, but one that may not have a satisfactory scientific answer. Why are customer inter-arrival times approximately memoryless? Why is locality of reference so prevalent? It is not difficult to advance philosophical ruminations, but perhaps few lead to scientific rigor or productive consequence. We leave the empirical laws as axioms.

4   POWER-LAW NETWORK TOPOLOGY

4.1   WHAT IS POWER-LAW NETWORK TOPOLOGY?

In the late 1990s, empirical measurements showed that a number of real-world graphs, including World Wide Web graphs [12, 22, 79], Internet domain net-

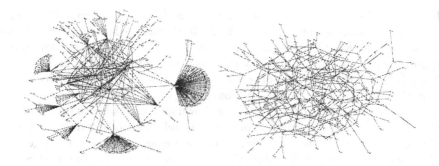

FIGURE 14   Power-law vs. random network topology. (a) 300-node Internet AS graph.
(b) 300-node random graph with the same edge density.

works [50], call graphs [1, 3], and certain biological and social networks [71, 97, 114] exhibited an unexpected connectivity pattern: the neighborhood size of nodes was highly variable, following a power-law distribution. As with heavy-tailed file sizes in self-similar traffic, this implied that most nodes are small, that is, have few neighbors, but a few are very large. The reason these findings were surprising is that they did not fit existing models—perhaps even conceptions—of how networks are connected, exemplified by random graphs whose size distribution has an exponentially decreasing tail. This does not mean that random graphs were regarded as good models of engineered and natural networks. In the field of combinatorial optimization there has been a perennial search for realistic benchmark graphs on which the average performance, as opposed to worst-case performance, of optimization algorithms could be evaluated. In spite of this recognized deficiency and need, pinning down the essential features of real-world networks proved elusive.

Figure 14(a) shows a 300-node Internet domain graph where nodes are administrative domains and edges denote peering relations between domains. Administrative domains, also called autonomous systems (ASes), are identified by 16-bit numbers (e.g., Purdue University has AS number 17). A typical peering relation is one where one domain is a customer of another, the provider. An AS is a logical entity that need not be geographically localized: for example, a major transit domain that provides connectivity service to other domains may have points-of-presence (POPs) across multiple continents where access routers are deployed. Access routers are connected by backbone networks internal to an AS. A link between two ASes means that there is at least one pair of border routers belonging to the respective domains that are directly connected, sometimes through a multiparty connection called an exchange. In an Internet domain graph, these and other details are ignored. Figure 14(b) shows a 300-node random graph with the same edge density as the 300-node Internet domain graph.

In $G_{n,p}$ random graphs—n is the number of nodes and p a probability—an instance of a random graph is generated by selecting each edge independently with probability p. Thus $G_{n,p}$ assigns a probability distribution over the finite sample space of all n-node graphs. If $p = 1/2$, all n-node graphs are equally likely. A graph is sparse if the number of edges is sub-quadratic, that is, significantly less than the total number of edges $\binom{n}{2}$. Power-law graphs turn out to be sparse. The connectivity structure of the two graphs shown in figure 14 is markedly different. Whereas the random graph is homogeneous and looks locally about the same, the power-law graph possesses "hubs" of varying sizes—some very large—that are connected through a "backbone." In a random graph, the neighborhood size of a node, called its degree, is concentrated around the mean np, a consequence of LLN. A random graph is an approximately regular graph, where a graph is k-regular if all its nodes have the same degree k. The graph in figure 14(a) is, in this sense, highly irregular and not a typical instance of $G_{n,p}$.

4.2   POWER-LAW RANDOM GRAPHS

4.2.1   Power-law Degree Distribution.
Random graphs, pioneered by Erdös and Rényi [47], possess a binomial degree distribution, thus making nodes with many neighbors exponentially rare. In a power-law graph, a polynomial degree distribution is postulated,

$$\Pr\{\deg(u) = k\} \propto k^{-\beta} ,$$

where $\beta > 0$ is an exponent that depends on the application area from whence a graph comes. For example, in a number of empirical measurement graphs, including Internet AS graphs, $2 < \beta < 3$. Figure 15(a) shows the degree distribution, on a log-log scale, for Internet AS topologies from Oregon Route-Views measurement data that are based on route table dumps [96, 135]. For a wide range of degree values—99% of nodes have degrees less than 50 and 95% have degrees less than 10—we observe a linear relationship with β slightly exceeding 2. For high-degree nodes whose small relative frequency is drowned out in a degree distribution plot, a rank distribution plot can be used where nodes are sorted in nonincreasing order of degree and the resultant rank is related to degree [50]. For example, in a 2002 Internet AS graph, rank 1 is occupied by AS 701—UUNET, a tier-1 transit provider—which has degree 2,538. By plotting rank versus degree, focus is shifted to high degree nodes at the expense of low degree nodes where an overwhelming majority have degree 1 or 2. Figure 15(b) shows the log-log rank distribution for the same data set. We observe a linear fit, consistent with a power-law relation, with slope a little less than 1. For a number of technical reasons, Internet AS topologies inferred from measurement data provide a partial and inexact view of domain-level connectivity. Although power-law degree distribution is a robust phenomenon observed across AS topologies obtained from different measurement sources, when drawing conclusions on the implications of power-law connectivity, it is imperative to consider the limitations

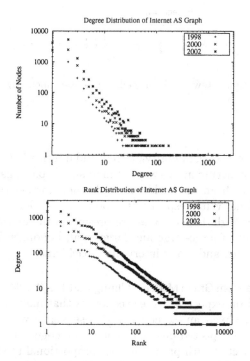

FIGURE 15 Power-law degree distribution of Internet AS graph. (a) Log-log degree distribution. (b) Log-log rank distribution.

of the measurement data and inferred topologies as well as application-specific idiosyncracies.

4.2.2 Power-law Molecular Stew.

From a structural perspective, we may interpret a degree sequence that specifies the degree of nodes in a graph in nonincreasing order as a set of n "molecules" where molecule i has w_i bonds. Figure 16 illustrates the ingredients of a power-law "molecular stew" which may be stirred to generate higher order structures. It is reminiscent of simulated annealing, albeit in multidimensional space as three-dimensional is too restrictive for effective stirring. High-degree nodes, by virtue of their abundant links, are likely to form connections with each other—not always directly—yielding a dense connected component comprising a skeleton backbone. Since high-degree nodes are few, many unfilled bonds will dangle from the backbone. The bulk of dangling bonds must link up with low-degree nodes, the most common building blocks in the power-law ingredient pool. This yields locally star-like shapes—reminiscent of Asian fans—a characteristic feature of power-law graph drawings (see, e.g., fig. 14(a)). The bondings, thus far, were of "large-large" and "large-small" kinds.

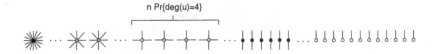

FIGURE 16   Molecular stew with ingredients determined by power-law degree sequence.

The remaining molecules—a minority of small and intermediate nodes—may contribute to the final structure in two ways: "small-small" pairings lead to elongated branches sticking out from the backbone, and others further the connectivity of the skeleton backbone. With respect to the power-law exponent β, the larger the exponent, the fewer high-degree nodes are present in the molecular stew, which diminishes the effect of large-large and large-small bondings. This may lead to fewer and smaller fans, and a less intertwined backbone.

4.2.3   Power-law Random Graph Model.   Chung and Lu [32, 84] studied a random graph model based on expected degree sequences that may be viewed as a generalization of $G_{n,p}$. Given an expected degree sequence $\mathbf{w} = (w_1, w_2, \ldots, w_n)$ where w_i denotes the expected degree of node i, an edge between nodes i and j is independently selected with probability p_{ij} proportional to the product of the weights, $p_{ij} = w_i w_j / \sum_{\ell=1}^n w_\ell$. It is easily verified that the expected degree of node i is w_i. $G_{n,p}$ may be viewed as the special case where $\mathbf{w} = (np, np, \ldots, np)$. The condition $\max_i w_i^2 < \sum_j w_j$ is placed which assures that p_{ij} is less than 1 and the degree sequence $\mathbf{w}$ is graphical, that is, there exists a graph with the given degree sequence. A related approach based on exact degree sequences, called a configuration model [92], has been investigated by Aiello et al. [3]. A key advantage of the expected degree sequence random graph model is its built-in independence. It comes, however, at the cost of defining a graph family whose members satisfy a prescribed degree sequence only on average. Erdös and Gallai [46] provide necessary and sufficient conditions for a degree sequence to be graphical, which also admits an iterative procedure for constructing a graph instance. We may generate other instances of graphs with a given degree sequence by constructing a Markov chain that performs a pair-wise edge switching operation: given two disjoint edges, the disconnected end points are joined and the original edges deleted. Clearly the local graph perturbation preserves degree sequence. In the configuration model, we may produce power-law multi-graphs—in a multi-graph two or more edges are allowed between a pair of nodes—by making w_i copies of each node i (each copy is void of edges), then forming a random matching on the resultant $\sum_\ell w_\ell$ nodes where nodes are randomly paired. When the w_i copies of node i are collapsed back into a single node, node i may share two or more links with another node j. Rigorous results can be shown for the configuration model with power-law degree sequence under random matching, such as the existence of a giant component and logarithmic bound on the size of smaller components [3]. Little is known about the Markov chain model.

In Chung and Lu [31, 32] and Lu [84] the basic properties of random graphs with a given expected degree sequence are established. Chung likens random graphs that obey a power-law expected degree sequence with $2 < \beta < 3$ to an octopus. The body of the octopus is a dense core of diameter $O(\log \log n)$ that contains $n^{c/\log \log n}$ nodes. The nodes in the core are large in the sense that their degrees are at least $n^{1/\log \log n}$. The average distance of smaller nodes to the core is $O(\log \log n)$. The octopus has arms that extend to $O(\log n)$ from the core. When $\beta > 3$, the power-law random graph is more expansive with average distance $O(\log n)$. At $\beta = 3$, the average distance is of order $\log n / \log \log n$. Some important features do not depend on the higher-order properties of a degree sequence. For example, the distribution of connected components, including the giant component, depends on the average (expected) degree $\sum_{\ell=1}^{n} w_\ell / n$ but not on the individual make-up of the weights. The small diameter—and even smaller average distance—is indicative of Milgram's small world phenomenon [91]. In social networks where links are defined by acquaintance relationships, six hops may suffice to reach presidents and Hollywood actors, a phenomenon referred to as "six degrees of separation." Since classical random graphs also possess a logarithmic diameter, the small world phenomenon—highlighted in the influential work of Watts and Strogatz [140]—is an important but perhaps not determining feature of empirical networks modeled by power-law graphs. In section 4.4 we consider more subtle properties with implications to shielding the Internet from network security attacks.

4.2.4 Growth Model: Preferential Attachment. A random graph evolution is a stochastic process that "grows" a graph instance by sequentially adding edges or nodes. It is a tool introduced by Erdös and Rényi [48] to study emergent structural properties of random graphs, including phase transitions at critical edge densities. In Barabási and Albert [12] a graph evolution for power-law graphs is proposed where a new node is preferentially attached to existing nodes based on their degree: a node is chosen for attachment with probability proportional to its degree. Thus a node with many connections is likely to get even larger as its chances of winning in the connect-to-the-new-node competition is self-reinforcing. This growth dynamic captures one form of the saying "the rich get richer and the poor get poorer," and may be a causal factor underlying the skewed connectivity of power-law graphs. In Bollobás et al. [18] it is shown that a formalization of this process leads to a random graph with a power-law degree distribution with exponent $\beta = 3$. In Lu [84] a generalized graph evolution is studied that is able to generate power-law graphs with a tunable exponent.

As a causal explanation of power-law graphs, preferential growth models are intuitively appealing since they embody biases in social dynamics where popularity, reputation, or notoriety have a tendency to become concentrated, at least up to a point. In the context of interdomain peering, organizational behavior—

other things being equal—may sway a new domain to subscribe bandwidth from a large, well-known transit AS due to: a tacit perception of reliability, simply because smaller providers are unknown, a bandwagon effect ("many of our business associates are connecting to UUNET, so should we"), and one-hop reachability to a large customer base. However, these are but a subset of relevant factors and other things are not always equal. For example, a medium-sized provider, due to political connections, may have POPs at international sites that a larger provider does not. In some instances, a smaller provider may have beaten a larger transit AS to the finish line with respect to new technology upgrade. Time scale is another important factor. Interdomain peering agreements occur at the time scale of weeks, months, and years, during which changes in economic climate, political upheaval, and technology innovation enter into the fray. Nonstationarity has been an issue when modeling self-similar traffic via stationary processes [44]: traffic at noon is different from traffic in the morning. Time-of-day periodicity, however, exists at time scales larger than those relevant for self-similar traffic engineering. For power-law topologies arising in application areas such as interdomain connectivity, router-level topology, call graphs, and Web graphs, nonstationarity due to exogenous variables may be more problematic when advancing causal explanations of power-law connectivity.

4.3 PERFORMANCE IMPLICATIONS AND CONTROL

4.3.1 Network Flow and Load Imbalance.

In this chapter, we focus on the implications of power-law connectivity when the underlying graph represents a flow network, as is the case for Internet AS and router graphs. Web graphs, in contrast, are first and foremost information networks where relational structure as captured by various criteria of semantic closeness, including the presence of clusters, admit analysis of massive data sets for effective information retrieval [10, 67]. For example, Google's Web page ranking algorithm [101] uses a popularity index—how many pages refer to a Web page—to determine the importance of search results. Flow and relational structure are intimately related, and some of the points advanced for flow networks have meaningful interpretations in the information retrieval context.

Let us consider a network $G = (V, E)$ where a routing algorithm has determined a path $u \rightsquigarrow v$ between every pair of nodes $u, v \in V$ in the network. Assuming uniform traffic demand, that is, all source-destination pairs are equally likely, we define the load of a node u, $L(u)$, as the number of paths that traverse through u. Thus the node load captures an innate "stress" that is placed on nodes in the flow network—a function of topology and routing—which can translate to hot spots or congestion if the load of a node is large compared to its capacity. An analogous definition holds for edge load $L(e)$, $e \in E$. Figure 17(a) shows the log-log plot of node load as a function of load rank—nodes are ordered with respect to their load where rank 1 is occupied by a maximal load node—for a 2002 Internet AS graph and a random graph of the same size and edge density.

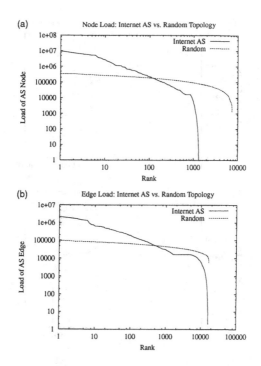

FIGURE 17   Comparison of load imbalance between Internet AS and random topology.
(a) Node load. (b) Edge load.

Figure 17(b) shows the corresponding log-log plot of edge load. For both Internet
AS and random graphs, we observe an approximately linear regime followed by
a sharp fall. The flat region in figure 17(b) toward the tail is due to ties in the
rank. A smaller flat region is discernible in figure 17(a). The main difference between Internet AS and random graphs is that the slope of the log-log plot in the
former is steeper than that in the latter. This implies that the load imbalance
in the Internet AS topology is more pronounced. In random graphs, routes are
dissipated and the resulting load is more balanced. Power-law connectivity has a
propensity to induce hot spots, skewing the responsibility placed on some nodes
and edges over others.

  This structural difference between Internet AS graphs and random graphs
is insensitive to details in the underlying routing. Figure 18(a) shows the log-log
node load distribution of the 2002 Internet AS topology under three different
routings: shortest-path, semi-random, and random. In shortest-path routing, a
shortest path is computed between source and destination using a destination-
based version of Dijkstra's algorithm, which iteratively constructs a routing tree

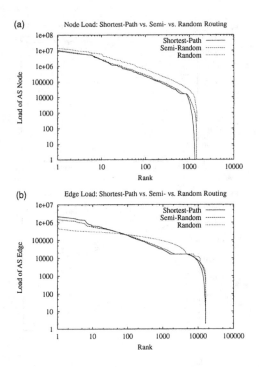

FIGURE 18 Impact of routing on Internet AS load imbalance: shortest-path, semi-random, and random routing. (a) Node load. (b) Edge load.

rooted at the destination. Destination-based route construction follows the procedure employed by BGP, Internet's global routing protocol. In semi-random routing, the shortest-path preference is modified such that when a new node is added to the routing tree—the purview of policy in interdomain routing—instead of choosing a minimal distance node, a random node is selected. The randomly selected node, however, is then attached to a node in the tree that it is closest to, preserving a tendency for inducing short paths. In random routing, both the next node selection and its attachment to the routing tree are done randomly. The average path lengths of shortest-path, semi-random, and random routing are 3.64, 3.83, and 6.57, respectively. The corresponding maximum path lengths are 11, 16, and 22. In figure 18(a) we observe that non-shortest-path routing has little influence on the shape of the node load curve. In power-law networks, imbalances in node load cannot be evened out by routing. The skewed node load distribution is an invariant of the power-law topology. Figure 18(b) shows that fully random route construction—an extreme form of policy diversity—does improve the balance in edge load. Semi-random routing has little effect on edge load. By forgoing route efficiency, it is possible to balance the edge load but not the node load ("all roads still lead to Rome").

The consequences of severe load imbalance on network engineering are: (1) a structural propensity for congestion if capacity is not accordingly matched, (2) congestion at a few nodes—node or edge failure being a special case where nothing passes through—impacts a significant fraction of the overall traffic demand, and (3) load imbalance is an innate property of power-law networks where network engineering exacts only limited influence. A caveat to the above is the assumption of uniform traffic demand which is not reflective of actual Internet demand. The Internet is principally a client/server network where demand is closer to "many-to-few" than "many-to-many." Voice-over-IP (VoIP) and peer-to-peer file sharing applications such as KaZaA promote a flat demand structure that may amplify in the future, but at the present they comprise a minority. When traffic demand is many-to-few, caching becomes an effective network engineering strategy that can keep a significant portion of the traffic demand local. One method for selecting cache proxies is to pick high-degree nodes where replicated files and services are hosted. High-degree nodes will continue to carry high processing burdens, however, their transit burdens will be significantly reduced. Topology-aware caching has been considered in Kamath et al. [73].

Multicasting is another application that can benefit from topology-aware placement. In multicasting, a spanning tree is constructed that allows broadcasting among a group of users. Multicasting on power-law topologies leads to graph embedding problems that aim to minimize various cost functions including the size of the multicast tree. Of related interest is the structure of multicast trees when the two-level hierarchy of Internet routing—interdomain at the AS level and intradomain at the router level—is incorporated. Assuming members of a multicast group are chosen randomly from the nodes in an Internet AS topology, the scale-free nature of power-law topology—random subgraphs retain their power laws with the same exponent—allows a scale-free characterization of the structure of AS level multicast trees once the multicast algorithm (for example, minimum spanning tree) is specified. Considering that transit domains are made up of router level backbone networks, if an AS level multicast tree is further expanded to include details at the router granularity, we arrive at a fine granular picture of multicast trees that incorporates the connectivity structure of router networks. Similar to path length inflation when routes are computed in a hierarchical fashion [130]—shortest-AS-path at the interdomain level followed by shortest-router-path within each domain—multicast trees computed in a two-level fashion are expected to be less efficient than those constructed from the global router level topology. Perhaps most intriguing is a succinct characterization of multicast trees obtained from two-level hierarchical routing, which may exhibit scaling properties such as those observed in Chuang and Sirbu [30]. Szpankowski et al. [128] have proposed a somewhat curious-looking tree structure, called a self-similar tree, that possesses distinctive unary segments in an attempt to explicate the power-law scaling behavior of multicast trees studied in Chuang and Sirbu [30] and Phillips et al. [111]. We conjecture that power-law connectivity, under two-level hierarchical routing and the existence of strong correlation between AS size and AS degree [129], induces multicast trees that resemble self-similar trees. Lastly, we note that for performance considerations in AS level graphs, "capacity," "congestion," "failure," and other performance measures must be carefully applied to yield meaningful interpretations. For example, a tier-1 transit AS is geographically dispersed with POPs across the continental United States and select international sites. A node failure in an Internet AS topology corresponding to such a transit AS would not be meaningful, even under massive power outages and other large-scale disturbances. This also holds for edge failures since a single AS level link between two transit domains may correspond to two or more physical peering points located in different parts of the country, say, one in New York, another in LA, and a third in Indianapolis. Unless all three physical connections go down, the two transit domains remain connected and their AS level link up.

4.3.2 Vertex Cover: Distributed Control and Optimization.

A vertex cover of a graph $G = (V, E)$ is a subset of vertices $S \subseteq V$ such that every edge $e = (u, v) \in E$ is incident on S, that is, either u or v (or both) belong to S. A vertex cover (VC) achieves a covering of edges in a flow network which allows distributed detection and control of communication events involving the generation and forwarding of packets. In the next section we will discuss its application—in the form of distributed packet filtering—to denial of service and worm attack prevention. In this section, we are interested in finding small VCs in Internet AS graphs. Since nodes in a VC are engaged in detection and control, having a small node set that covers all edges, that is, is a VC, is important for economy and ease-of-deployment. For example, in the global Internet, getting one domain to adopt a new technology is a complicated matter. To get multiple domains to agree on a common technology base, such as the deployment of distributed filters, is an even tougher challenge due to administrative autonomy, policy barriers, and conflicts of interest. Thus, our primary cost measure with respect to implementing distributed detection and control is the size of a node set—VC or otherwise—where actions are undertaken.

Finding a minimal VC in an arbitrary graph is an NP-complete optimization problem [58]. A problem is in NP if checking whether a candidate solution is indeed a solution is easy, that is, it can be done in polynomially many steps in the size of the input. Finding a solution may be easy—then the problem belongs to the class $P \subseteq NP$—or it may not. Intuitively, we would think that appreciating good music is easier than composing it. Artists and scientists would not be able to make a living without this maxim. Critics are professional solution checkers. The "P vs. NP" problem in computer science captures this popular and, on the surface, obvious truth but, to date, no one has been able to prove that $P \subsetneq NP$, that is, the existence of a problem for which the solutions are easy to check but difficult to find. A problem is NP-complete if it is in NP and dominant: the ability to solve it would make every other problem in NP solvable. Finding a minimal VC is such a problem. Although a fast algorithm for minimal VC is not known—likely because none exists—there is a fast algorithm that always gives a VC that is at most twice as large as a minimal one. The algorithm starts from an empty cover, iteratively picks an edge, inserts the two incident nodes into the cover, removes the two nodes and any edges incident on them from the graph, and repeats the process on the smaller graph until no more edges remain. It is easily verified that the algorithm—more accurately the procedure, since we haven't specified how to pick the edges—achieves a factor 2 performance guarantee. A drawback of this algorithm under random edge selection is that it does not do well in practice: it is outperformed by a greedy algorithm that grows a VC by first inserting a node with the highest degree, removes the node and its incident edges from the graph, and repeats the process until no more nodes remain. An issue with the greedy algorithm is that no rigorous performance traits are known except that there are graphs on which the greedy heuristic fares poorly: the size of the VC found is at least a factor of $\log n$ bigger than a minimal VC. In the following, we show the size of small vertex covers found by running both the greedy heuristic and factor 2 approximation algorithm. The latter chooses edges randomly in the iterative growth process, and for each graph instance is run 100 times with different random seeds. For the graphs benchmarked below, the greedy algorithm always found smaller VCs compared to the factor 2 approximation algorithm. The factor 2 approximation algorithm, however, is useful for lower-bounding the minimum VC size. From 100 runs, pick the worst VC size found, divide it by 2, and it gives a lower bound on the minimal VC size.

Figure 19(a) shows the size of Internet AS topologies inferred from NLANR/ Oregon Route-Views measurement data for the period 1998–2002. We observe a slight super-linear trend. Figure 19(b) shows the smallest VC size found for the Internet AS topologies and corresponding random topologies of the same size and edge density. The VC size for Internet AS graphs falls below 15% in 2002, whereas for random graphs it increases above 55%. A practical implication of the VC gap is that if Internet AS connectivity were random, significantly more nodes would be needed to cover all edges in the network. Power-law connectivity affords economy of deployment through strategic placement. To ascertain the accuracy of the estimated minimal VC sizes, we plot both the upper bound (obtained by the greedy algorithm) and lower bound (obtained with the help of the factor 2 approximation algorithm) in figure 20(a). We observe that for Internet AS topologies the two bounds are very close, achieving an accuracy level greater than 90% as shown in figure 20(b). In the case of random topologies, the upper and lower bounds are significantly further apart leading to an accuracy level below 75% in 2002. Asymptotic estimates for the size of a minimal VC are known for $G_{n,p}$ random graphs under the condition that the average degree be large (i.e., $np \rightarrow \infty$) [56]. Since power-law graphs possess a constant average degree, the corresponding random graphs must have a constant average degree to achieve the same edge density. One of the technical challenges in power-law

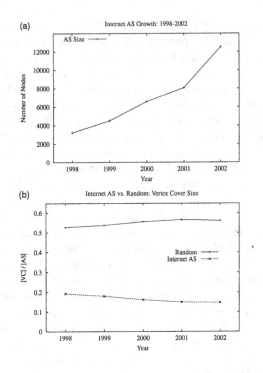

FIGURE 19   Internet AS topology. (a) Growth during 1998–2002. (b) Vertex cover size.

graphs is to understand what implications power-law connectivity has on combinatorial optimization. The presence of large degree nodes that induce locally star-like subgraphs makes optimization, to a first approximation, easier. However, an additional increase in accuracy requires an improved understanding of the "backbone" structure of power-law topologies where intermediate-sized nodes play an important role. VC is a good starting point to study these questions as finding a minimum VC is intimately related to finding a maximum independent set and clique. An independent set of a graph $G = (V, E)$ is a subset of vertices where the nodes are mutually disconnected. A subset of nodes is a clique if everyone is connected to everyone else. It can be easily checked that if $S \subseteq V$ is a VC, then $V \setminus S$ is an independent set of G and a clique of $G^c = (V, E^c)$ where E^c is the complement set of E. Another interesting avenue for optimization in power-law graphs is a deterministic definition of power-law graphs on which optimization questions are studied. For example, given a (graphical) degree sequence $\mathbf{w} = (w_1, w_2, \dots, w_n)$, the set of all graphs that satisfy the degree sequence is well-defined. One would like to define a sequence of graph families $\mathcal{G}(n, \beta, r)$ where β is the power-law exponent and r a fudge factor—G is allowed

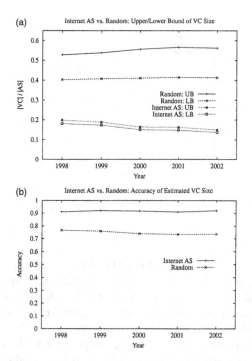

FIGURE 20   Accuracy of estimated minimal VC size. (a) Upper and lower bound of minimal VC size for Internet AS and random graphs. (b) Accuracy of estimated minimal VC size.

to be a member of $\mathcal{G}(n, \beta, r)$ if its degree sequence is r-close to the power-law degree sequence—where optimization can be addressed in a deterministic setting. Does minimum vertex cover remain NP-complete for $\mathcal{G}(n, \beta, r)$? One suspects that this is still the case due to the combinatorial messiness that the "backbone" of power-law topologies seems to harbor, wherein smaller—but polynomially sized—hard problems may be embedded. Finding a robust formulation of deterministic power-law graphs that can be related to the average degree sequence framework and is amenable to analysis is a first challenge.

We conclude this section with a note on artificial power-law topology generation that emulates real-world graphs, in particular, Internet AS topologies. Figure 21 shows the VC sizes for artificial topologies produced by Inet [72, 143], a topology generator that aims to mimic empirical Internet AS connectivity. There is a significant discrepancy in VC sizes between Internet AS topologies and corresponding artificial topologies generated by Inet-2.2 [72], which was first pointed out in Park and Lee [103]. This has, in part, prompted modifications to Inet-2.2 leading to Inet-3.0 [143], which yields topologies with VC sizes close

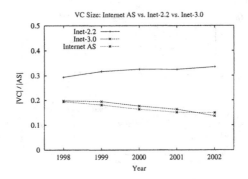

FIGURE 21   VC size comparison: Internet AS vs. Inet-2.2 vs. Inet-3.0.

to that of Internet AS topologies. One of the key changes in Inet-3.0 is the incorporation of "higher order" connectivity structure, meaning that in addition to how many neighbors a node possesses, its neighborhood composition with respect to degree distribution is injected into the graph construction process. The performance of Inet-3.0 has significantly improved over that of Inet-2.2, however, there remains room for improvement, especially with respect to clustering [143]. A topology generation approach that captures higher order connectivity structure—$A, A^2, A^3, \ldots$ where $A = (a_{ij})$ is the adjacency matrix of a graph, that is, $a_{ij} = 1$ if i and j are connected, 0, otherwise—may harness additional structure that is presently missed. As with second-order stationary processes for modeling self-similar traffic, higher order structure beyond a certain point may not matter.

4.4   APPLICATION: DENIAL OF SERVICE AND WORM ATTACK PREVENTION

4.4.1   What are Denial of Service and Worm Attacks?   A denial of service (DoS) attack aims to disrupt services by depleting network resources such as bandwidth and CPU required to deliver the services. This is done by sending bogus work in the form of junk traffic and service requests that tie up resources, preventing a network system from operating in its normal mode. Attacks may be targeted at servers, hosts, and increasingly at the network infrastructure itself [93]—such as routers and name servers—with far-reaching impact. Distributed denial of service (DDoS) attacks that forge their source IP address, called spoofing, are especially severe, due to their concentrated force and difficulty in affecting timely recovery. Locating an attack source can take on the order of hours, sometimes days, by which time the damage has already been done. Vulnerability to denial of service is not unique to the Internet. An old form of DoS attack is "ring the door bell and run" practiced by select kids in days gone by. Although innocuous,

the occupant's time and energy are wasted, drawing attention away from other matters. On the Internet, repercussions from DoS attack are amplified due to the high attack volume enabled by broadband networks, ease of anonymity afforded by spoofing, and fingertip attack coordination from a computer anywhere in the world. Classical DoS attacks are many-to-one in the sense that an attacker recruits multiple attack hosts to overwhelm a target. Denial of service from spam mail, on the other hand, is one-to-many which is made possible by the slow processing speed of the target—human leafing through e-mail—and the financial incentives underlying the denial of service activity. One of the few effective defenses against DDoS attacks is assigning a cost to the attacker, either in the form of economic cost (e.g., usage pricing) or attacker identification followed by attribution with legal ramifications. Both achieve a deterrent effect, a principal reason why a DoS attacker wouldn't buy up all the baguettes at the local bakery, or run off without paying.

A worm is a computer virus that travels from system to system on the Internet through network protocols invoked by distributed applications. A computer virus is parasite code—a block of instructions and data embedded in application or system code, its host environment—that upon execution can impart harm to a computing system through information erasure, system hijacking (e.g., for the purpose of DoS attack), infection, and self-replication, among an array of possibilities. In the 1980s, a preferred way for viruses to infect other systems was through floppy disks that contained infected applications or had infected boot sectors. Today's mobile viruses—worms—exist as parasitic attachments to messages that are transmitted, interpreted, and processed by network protocols as part of networked client/server applications. Worms face a tougher challenge than their brethren when aiming to take over a system. They need to remain clandestine—network protocols, which are structured programs expecting well-formed messages, must not see through the disguise—until they reach a specific point in the protocol code that contains a fatal vulnerability. The message containing the worm is structured so that it triggers the vulnerability leading to a coup d'état. The most prevalent vulnerability targeted by past worms, including the Morris worm (also called the Internet worm [127]) and Code Red [26], is the buffer overflow vulnerability. When a message is parsed, arguments are copied into the program's memory space holding data. If an argument is overly long, it may spill over into the program's instruction space, overwriting the program's instructions with parasite code hidden in the argument. To pull this off requires expertise on the part of an attacker, but the vulnerability would have been prevented had the code been more carefully written to check the length of arguments.

4.4.2 Fixing a Hole in the Security Roof.

Before we discuss the role played by power-law connectivity in "fixing a hole where the rain gets in" (to paraphrase Lennon and McCartney), let us examine some pertinent features of DDoS and worm attacks that constrain the types of defenses that may be mounted against them. An unspoofed DDoS attack is inherently difficult to shield against. Stemming an unusually large traffic flow—if deemed a DDoS attack—is an application of flow control and not difficult to do. The crux of the problem is: how does one distinguish friend from foe? There are no known methods for keeping false positives in check, ensuring that "the baby is not thrown out with the bath water." The Achilles' heel of intrusion detection systems is that anomalies, statistical or otherwise, are not sufficiently reliable to warrant automated responses that may, in the end, do more harm than good. They are useful to detect increased "chatter," but determining what the chatter means is, at the present, more art than science. If DDoS attacks are unspoofed, then their IP source addresses reveal the identity of the traffic source—attacker and legitimate—which then becomes the purview of policy actions. Economic factors such as usage pricing—in the U.S. most Internet access is based on flat pricing—have only a small bearing on the problem as attackers, who compromise others' machines, do not bear the economic burden. A meticulous attacker who recruits hundreds of attack hosts by planting viruses well in advance of a timed DDoS attack is difficult to catch. A less meticulous attacker who has not sufficiently hidden his/her tracks—for example, by supervising an unspoofed DDoS attack in real-time—is much easier to apprehend. An effective solution, and perhaps the only long-term solution, for protecting against DDoS attacks is deterrence through attribution, whose first step involves timely source identification.

Worm propagation shares many similarities with virus propagation in biological systems, the subject matter of epidemiology [11, 64]. Both are contact processes that spread through specific interaction, and in the absence of counteracting forces, rapidly infect the bulk of a susceptible population. The mathematical foundations of epidemiology are twofold: a macroscopic theory of population dynamics based on dynamical systems theory, and a microscopic theory of disease propagation based on percolation theory. The classic SIR epidemic model [64], which has its origins in the seminal work of Kermack and McKendrick [75], relates susceptibles (S), infectives (I), and removed (R)—that is, the dead or otherwise inert—in a population N (where $N = S + I + R$) through a system of nonlinear differential equations. It is assumed that dI/dt increases proportional to SI but decreases proportional to I, $dS/dt \propto -SI$, and $dR/dt \propto I$. The product form SI captures the uniformity assumption that each susceptible is potentially exposed to a fixed fraction of infectives (and vice versa). Ignoring the dead—worms utilize infected systems for further propagation—the transient behavior of the population dynamics exhibits a characteristic "S-curve" where all susceptibles eventually get infected. With R in the picture, the steady state population may contain survivors, that is, non-zero susceptibles, when the death rate exceeds the infection rate which drives I to zero. In an endemic model, death is accompanied by birth which, in the worm propagation context, may be interpreted as infectives being healed by reinstalling or cleaning up of infected systems. An S-curve characterizes the spread of worms on the Internet, exemplified by Code Red whose spread was monitored by CAIDA [23] and shown at a DARPA meeting in July 2001 amidst a sequence of attacks.

4.4.3 Distributed Packet Filtering. Cooperative Protection and Partial Deployment.

Distributed packet filtering (DPF) is a new approach to protecting the Internet against DDoS and worm attacks. DPF embodies a "cooperative protection under partial deployment" paradigm that breaks with tradition in two important respects:

- *Local vs. global protection.* Firewalls epitomize local protection whose aim is to shield an entity from adverse outside effects. Amazon.com may succeed in shielding itself against a worm attack; however, if a significant fraction of its customer base is disabled by the same attack, the impact is ultimately shared. Selfish protection—manifested as local protection—only goes so far.
- *Partial vs. full deployment.* A network security solution, to be effective, must consider partial deployment a fundamental maxim of Internet vulnerability. Epidemiology teaches us why software patches, when partially deployed, are unable to contain the spread of worms. Insisting on increased user diligence assumes the problem away, is unrealistic, and is not the domain of science and engineering.

The limitation of local protection under partial deployment exposes a locally protected service provider to DDoS attack: when others are unprotected, this allows them to be recruited for DDoS attack which turns others' vulnerability into one's own. A firewall suffers under lack of efficient friend-from-foe discrimination capability which impedes selective admittance of non-attack traffic. For DDoS and worm attacks, the fates of the protected and unprotected are intertwined. To overcome the limitation of local protection under partial deployment, a new dictum is required: (i) Protective action must be centered at transit points, not end systems, to exploit checkpoint screening and containment afforded by transportation networks. (ii) Transit points must cooperate to affect global protection; a locally protected network system is inherently vulnerable to DDoS and worm attack. (iii) The collective action of a few, under partial deployment, must yield an overwhelming synergistic effect that protects the whole. In general, realizing a solution that satisfies all three conditions is a tall order. The first two properties are architectural features and, as such, are part of the "cooperative protection under partial deployment" design space. Their viability critically depends on (iii)—a performance feature—and the most difficult part of the new approach. It is here where power-law connectivity plays a crucial role.

Casting a net over Lake Wobegon. To illustrate the notion of distributed packet filtering, suppose Lake Wobegon is being polluted by hostile elements carrying out water contamination attacks. Contaminated water affects fish and wildlife, and eventually threatens water supplies. Local filtering can cordon off a shore segment and purify the water therein for human consumption. However, inter-city commerce—mediated by ships and boats on water routes—dwindles for fear of admitting contaminated water particles. The fishing and sea food industry is shot. Cities that did not heed precautions turn into ghost towns. Under distributed filtering, a number of cities band together and install water filters across the whole lake. Individual filters are not aimed at protecting a specific town or city but the system as a whole. Any one community has little incentive to install a filter outside its immediate living space: a filter or two in the middle of the lake would not do much good anyway. It is only when a sufficient number is deployed that contamination introduced anywhere on the lake gets trapped by the filter net and further spread is contained. The state-of-affairs in Lake Wobegon before and after distributed filtering is illustrated in figure 22(a).

Distributed packet filtering only satisfies properties (i) and (ii): cooperative protective action carried out at transit points for the good of the whole. Supposing Lake Wobegon is "super-sized" to Internet scale, without property (iii) that mandates a small filter deployment with a big bite, distributed filtering would not be feasible. Only when an economic filter placement achieves decisive protection can relevant transit parties be brought together—some screaming and hollering may be involved—and induced to form a coalition for the greater good. The larger the required coalition, the less chance this has of succeeding. Thus, minimizing deployment becomes a key objective. In a transportation network, the paths that an entity can take are constrained by the underlying connectivity structure. In communication networks, routing further restricts the route that a packet can take. In road systems, checkpoints at border crossings and intersections have been used to shield countries and cities from unauthorized access. These gateways, when engaged as distributed filters, have been less effective at apprehending fugitives of justice: under partial deployment, there are simply too many ways for a person to travel from one point to another undetected. In power-law networks, the tables are turned. Even though checkpoints are few, it is next to impossible to go anywhere without encountering a sentinel. Figure 22(b) shows a power-law network with strategically placed distributed filters. In the next two sections, we will outline two forms of distributed packet filtering—route-based and content-based filtering—and summarize how global protection is effected by power-law connectivity under partial deployment.

4.4.4   DDoS Attack: What is Route-Based Filtering?

The primary task of a router, upon seeing a newly arriving packet, is to ask where it is headed (quo vadis) and send the packet on its way based on the answer provided in the IP destination address. We advance that a router ask a second question—"where do you hail from?"—and discard the packet if it can be unequivocally determined that it is lying. That is, the IP source address is spoofed. For example, if a packet arrives at an interface claiming to originate from A destined for B when, based on routing, it is impossible for such a packet to enter through the said interface, the packet is spoofing and dropped. When this verification is done without utilizing the

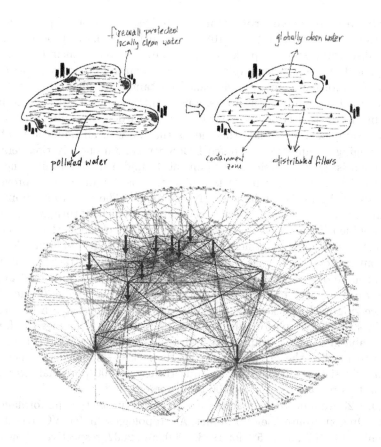

FIGURE 22 (a) Pollution in Lake Wobegon before, and after, distributed filtering. (b) Filter net for DPF in power-law network.

destination address—in the above scenario there may not exist any destination address for which a packet from A would enter through the given interface—we call it semi-maximal route-based filtering, in contrast to maximal filtering where both addresses are used. If, by exploiting route constraints, we discard spoofed packets before they can reach their target and impart harm—for example, as part of a spoofed DDoS attack—we endow the system with authentication capability without engaging cryptography. There are legitimate instances where spoofing is useful, such as in Mobile IP where an intermediary forwards packets on behalf of a mobile user. These protocols are not widely used, however, and can be modified to work without resorting to source address spoofing.

Proactive and reactive protection. Route-based distributed packet filtering was introduced in Park and Lee [103]. Its goal is to achieve proactive protection by discarding spoofed packets before they can reach their target, which stops spoofed DDoS attacks in their tracks. We quantify proactive protection by defining a performance measure Φ, called containment, where $0 \leq \Phi \leq 1$ denotes the fraction of innocuous nodes: a node is innocuous if any spoofed packet emanating from it, destined to anywhere, is discarded by a filter in the filter net. For example, if $\Phi = 0.95$, this means that 95% of all nodes are not suitable as staging grounds for a spoofed DDoS attack. An optimization version of the problem is: given a desired containment Φ, find a minimum filter net such that proactive protection with Φ is achieved. It is not difficult to prove that optimal spoofed DDoS containment is NP-complete. The need to minimize filter deployment may necessitate imperfect containment which, in turn, brings out the problem of identifying the source of an attack. In January 2002, the Internet had more than 12,000 domains, which leaves open the possibility that with $\Phi = 0.95$, hosts from 600 domains could collude in a spoofed DDoS attack. Locating the physical source of a spoofed IP packet, also called traceback, becomes an issue. We quantify traceback by defining k-traceability where a node is k-traceable if, upon receiving an IP packet, the physical source of the packet can be localized to within k sites. The performance measure $0 \leq \Psi(k) \leq 1$ defines the fraction of k-traceable nodes. For example, if $\Psi(6) = 1$, this would mean that all nodes are able to localize the origin of received IP packets—spoofed or otherwise— to within six sites. Compared to the uncertainty of 600 potential sites under $\Phi = 0.95$, this represents a factor-100 improvement.

Figure 23 summarizes proactive and reactive protective performance for NLANR/Oregon Route-Views Internet AS topologies under VC based filter placement: 19%, 16%, and 15% for 1998, 2000, and 2002, respectively. The graph size during the same period increased from 3,015 to 12,517. Figure 23(a) shows containment Φ for the three years which increased from 98% in 1998 to 99% in 2002. Thus with 15% filter deployment, we are able to achieve 99% proactive protection restricting the attacker to just 1% of the domains for staging spoofed DDoS attacks. Figure 23(b) shows the fraction of k-traceable nodes, $\Psi(k)$, as a function of k for the same benchmark topologies. We observe that for all three years $\Psi(4) = 1$. That is, all nodes, upon receiving an IP packet, can locate the origin of the packet to within four sites. In the 2002 topology, 85% of the nodes can narrow the range down to two sites. Constant containment and traceback performance in the presence of a fourfold increase in the size of the topology makes route-based DPF a scalable DDoS solution.

Robustness. As noted earlier, Internet AS topology data represent an approximate picture of AS level connectivity, which are imbued with inaccuracies introduced by the measurement data as well as inferences drawn from them. To ascertain whether route-based DPF performance is sensitive to the details of the underlying measurement data, we evaluate VC size, proactive and reactive

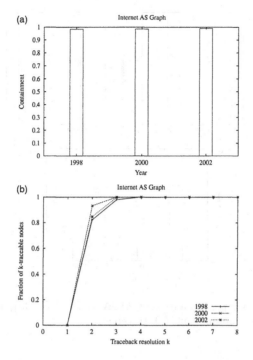

FIGURE 23   Spoofed DDoS attack protection. (a) Containment. (b) Traceback.

protection for a range of measurement topologies that draw on different measurement methods, data sources, and inference procedures. Figure 24 shows proactive and reactive performance of route-based DPF on Internet AS topologies from CAIDA [24], RIPE [120], USC/ISI [89], and the University of Michigan [134]. Figure 24(a) depicts VC sizes found by the greedy algorithm which fall in the 12–17% range for the four data sets. The VC sizes are, overall, consistent with those found for NLANR/Route-Views topologies. Figure 24(b) shows that containment lies in the 98–99% range, and Figure 24(c) shows the minimum value of k for which $\Psi(k) = 1$, which yields $k = 3$ or 4. That is, all nodes are able to localize the physical source of received packets to within three or four sites.

Random vs. power-law connectivity. Figure 25 shows the performance of route-based DPF on random graphs of the same size and edge density as the NLANR/Route-Views AS topologies. Figure 25(a) shows the much larger VC sizes found in random graphs (cf. also fig. 20(a)). For example, in 2002 the size of the VC filter net exceeds 55%. Figure 25(b) shows that despite the nearly four-times larger filter net, proactive protection in random graphs is significantly

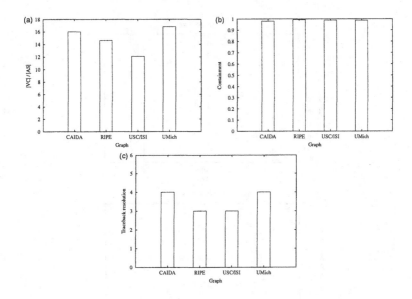

FIGURE 24   Protective performance on CAIDA, RIPE, USC/ISI, and UMich topologies. (a) VC size. (b) Containment. (c) Traceback.

reduced from 99% to 65%. Figure 25(c) shows that traceback performance has degraded as well, with the smallest k such that $\Psi(k) = 1$ reaching 27 in the 2002 graph. The results indicate that route-based DPF in randomly connected networks would not be a viable solution. Power-law connectivity is essential to scalable protective performance effected by strategic, economic filter placement.

VC filter placement and power-law connectivity. The vertex cover heuristic for selecting filter sites induces two properties relevant for facilitating proactive and reactive protection in power-law networks:

- *Preference of high degree nodes.* The greedy VC heuristic assigns preference to high degree nodes which leads to economy in filter deployment due to the prevalence of stub ASes that are connected to large transit ASes. For example, in the 2002 Internet AS topology, more than 1/3 of the nodes are stub domains of degree 1.
- *Uniform filter density along routes.* A more subtle property—critical for reactive protection—is uniform filter density along any end-to-end path. This is a consequence of the VC property: to cover all edges along a route, at least every other node must belong to the filter net. Uniform filter density severely limits "holes" through which spoofed packets can sneak in.

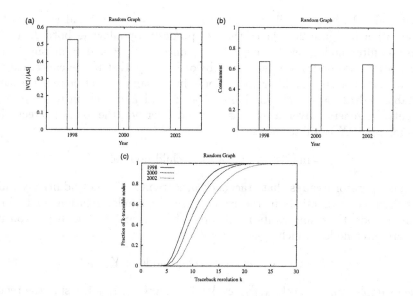

FIGURE 25 Protective performance of route-based DPF on random graphs. (a) VC size. (b) Containment. (c) Traceback.

Containment is primarily effected by the first property. Traceback is effected by the second property. A filter placement method that chooses the top 15% of high-degree nodes—a purified form that implements the first property—achieves traceback resolution 263, that is, $\Psi(263) = 1$, which is significantly worse than $\Psi(4) = 1$ achieved by 15% VC based placement. Moreover, $\Psi(261) = 0$. That is, no node achieves 261-traceability and there is a sharp transition at $k = 263$ ($\Psi(k) = 0.02$ at $k = 262$). The influence of the second property is illustrated in figure 26 which depicts a routing tree rooted at node J, the target of a spoofed DDoS attack. Assuming H is a non-filter transit node, by the VC property, A, D, G, and I must be filter nodes. They form a perimeter around H isolating it from the rest of the network. Let ψ_X denote the size of the maximum spoofable address space that a node in the subtree rooted at X can engage to attack node J. That is, X can craft a spoofed packet that reaches J—that is, not discarded by the filter net—and J cannot resolve the origin of the packet beyond ψ_X candidate sites. We assume filter nodes do not allow spoofed packets to emanate from within. We use ψ_J to denote the traceback resolution of J. If Y is the parent of X, then $\psi_X \leq \psi_Y$. Thus $\psi_J = \arg\min_k[\Psi(k) = 1] = \max_X \psi_X$ where $\Psi(k)$ is restricted to target J and the maximum ranges over all X in the routing tree of J.

In the example shown in figure 26, B is a stub node, a degenerate singleton subtree. Its parent D is a filter node, hence $\psi_B = 1$. The same holds for C. For transit node D which is a filter node, $\psi_D = 1$. $\psi_E = \psi_F = 2$ and $\psi_G = 2$. The configuration in figure 26 relaxes the VC property by allowing both E and F to be non-filter nodes, leaving edge (E, F) uncovered. E and F belong to the same pocket: two nodes are defined to belong to a pocket if one can reach the other without traversing a filter node. Nodes in the same pocket possess the same spoofable address space: F can claim to be E, and E can claim to be F, with impunity. In general, given a routing tree and filter net, the following recursion holds for ψ_X if X is a filter node:

$$\psi_X = \max\{\psi_{c(X)} : c(X) \text{ is a child node of } X\}.$$

The max operator assures that traceback uncertainty does not additively build up at filter node junctions in the routing tree. A similar relation holds for a non-filter node Y as long as its neighbors—children and parent in the routing tree—are filter nodes, which is assured by VC:

$$\psi_Y = \max\{\psi_{c(Y)} : c(Y) \text{ is a child node of } Y\} + 1.$$

In the example, $\psi_H = \max\{\psi_A, \psi_D, \psi_G\} + 1 = \max\{\psi_A, 2\} + 1$. Using the recursions, we can establish an upper bound on the traceback resolution when filter placement is VC: half the diameter of the given network. The exact bound is given by the maximum number of non-filter nodes on any path from stub to destination in the routing tree. Since power-law graphs have small diameter—for example, the 2002 Internet AS topology has diameter 11 (random power-law graphs have diameter $O(\log n)$)—this yields an analytical bound on traceback resolution in power-law graphs under VC filter placement that utilizes the small world property. The preceding development also shows how to relax the VC property to further reduce filter deployment: find a filter placement that yields small pockets so that the coarsified routing tree—nodes in the same pocket are grouped into a single (weighted) super-node—has accurate traceback resolution.

4.4.5 Worm Attack: What is Content-Based Distributed Packet Filtering?

In content-based distributed packet filtering, content-based filters are installed at strategic locations in a network such that packets carrying worm parasites are detected and discarded before they can reach their targets. Content-based DPF is both a generalization and specialization of content-based filtering carried out in firewalls. It is a generalization in the sense that filter deployment at transit points is guided by the principle of protecting the system as a whole, in contrast to local protection effected by firewalls. It is a specialization in the sense that content-based filters are primed to detect packets carrying worms, an undertaking whose scope is narrower and more focused than devising content-based filters in general. For example, in the Code Red worm, a cleverly formatted GET request—a type of HTTP message—along with a disallowed body ends up triggering a buffer overflow that ultimately transfers control to the worm code [88]. One simple filter rule is to discard any IP packet that transports a TCP packet carrying a

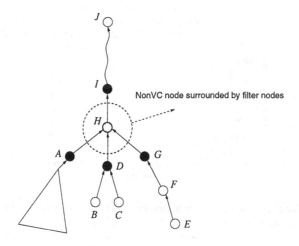

FIGURE 26 Routing tree rooted at node J. A, D, G, I are VC filter nodes, and H is a non-filter node.

HTTP GET message with a body: the HTTP standard stipulates that a GET message cannot have a body. The length of the body—3,569 bytes—prompts the malformed HTTP GET message to be split across multiple IP packets which can complicate the worm detection process at routers. In other content-based filters, the GET request may be inspected to ascertain if it is a Code Red worm—the request part contains a characteristic signature—without having to check for the presence of a body. Both filter rules are resistant to mutations where parts of the request or body are perturbed to defeat detection. This resilience in worm filtering is due to the structured nature of network protocols: worms must abide by protocol rules until such time as a target vulnerability—for example, a buffer overflow—can be triggered. This makes parasitic variants stick out, rendering them efficiently detectable ("it's difficult to conceal a gun, samurai sword, or other weaponry when doing a triple somersault"). A caveat is that recognition of a specific vulnerability—such as a buffer overflow in Windows indexing service IIS which Code Red exploits—must precede its prevention through content-based DPF. Anticipating and protecting against future instances of buffer overflow vulnerabilities is an impossible task: the vulnerabilities can be very subtle, as is the case in Code Red, instigated by an unbounded "steadfastness" exhibited by programmers adept at writing potentially buggy code. The championship belt in mediocre craftsmanship is held by a company in Redmond that produces equivalents of Trabants in the days of the iron curtain, but the responsibility is shared by many including computer science departments at universities where many of the engineers have been educated. The good news is that new worms such as Code Red were unleashed after public announcement of new vulnerabilities and

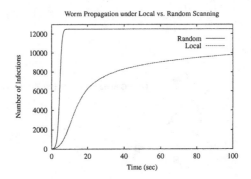

FIGURE 27 Worm propagation dynamics in 12,512-node Internet AS graph under local vs. random scanning.

patches. Deployment of updated content-based filters at distributed filter sites before public release can achieve preventative protection.

Worm propagation dynamics. Epidemiology teaches us that patching, due to partial deployment and local protection, cannot protect the segment of the Internet in which patch updates have not been installed. Persistence of partial deployment is also the root cause of why repeat attacks—many worm attacks are reincarnations—continue to wreak havoc weeks and months after first impact. Figure 27 shows worm propagation dynamics under two extreme contact—that is, victim selection—rules: local where nearest neighbors are selected, and random (or global) where targets are chosen randomly from a global address space. A lightweight preprocessing step, called scanning, precedes the actual worm transfer, aimed at discerning whether a target is potentially vulnerable. In actual attacks, both selection strategies, their mixture, and custom selection rules are used. At a scan rate of a few scans per second, it takes but a minute to infect a large chunk of the Internet. Containing worm attacks under partial deployment is only feasible by deploying filters at strategic transit nodes—part of a carefully managed infrastructure defense—through which IP traffic must pass through. When the underlying network topology is power law, the contact rule implemented by the application layer worm propagation has little effect. For example, in the Melissa worm [25]—called a macro virus because it is propagated as a Word attachment in Outlook e-mail that executes embedded code, in other words, macros, when opened—the contact rule consults the user's address book to determine subsequent targets. Whatever the connectivity structure induced by users' address books—perhaps a small world social network not unlike Milgram's [91]—power-law AS topology constrains the paths that a packet can take from source to destination, making it difficult for worms to travel undetected.

A defense architecture that deploys filters in mail server networks would not be viable due to its mesh connectivity: a mail message sent from a host in domain A to a host in domain B is, at its core, a transaction between mail servers at A and B. Mail servers at other domains do not get involved. Partial filter deployment in mesh networks only yields local protection, with the same performance limitation as firewalls.

Worm containment: power law vs. random connectivity. Given a network $G = (V, E)$, routing, and filter net $S \subseteq V$, v is reachable from u if the path from u to v does not contain a filter node. Thus a packet from u to v can travel undetected and is infectable by u, denoted $u \twoheadrightarrow v$. The relation "$\twoheadrightarrow$" is transitive and partitions V into equivalence classes where u and v are in the same equivalence class if both $u \twoheadrightarrow v$ and $v \twoheadrightarrow u$. "$\twoheadrightarrow$" also induces a partial order—more precisely, a forest of disjoint trees—on the resulting equivalence classes. We are interested in finding a small filter net S such that all trees are small. Since a single infected node at the root of a tree can infect all other nodes in the tree—the most economic way for a worm attacker to wreak maximum havoc—the smaller the trees, the more effort is required on the part of the attacker to contaminate a network system. If T is a largest tree, then at least $|V|/|T|$ nodes must be separately compromised to infect the whole network G. Conversely, a local outbreak in T is contained within T, the goal of isolation in disease control.

Figure 28(a) shows the size of a largest tree—we also call trees "pockets"—for the 2002 Internet AS topology under a pruned VC filter placement: $x\%$ filter density means that after computing a VC it is pruned to the smaller target size x by discarding low-degree nodes first. We observe that there is a critical filter density around 3–4% at which there is a sharp change in the size of a maximal pocket. The threshold phenomenon is also present in a corresponding random graph of the same size and edge density, albeit with the threshold located near 28%. The engineering implications are twofold: one, worm containment under economic deployment is facilitated by power-law connectivity, and two, deploying filters below the critical threshold is of little use, while deploying filters above the critical threshold is wasteful—knowledge of the critical filter density is crucial to achieving effective protection. Figure 28(b) shows the size of a second largest, third largest, fourth largest, and fifth largest pocket under different filter densities. We observe that containment is dominated by the largest pocket, which is unique. Figure 29(a) depicts the distribution of pockets in a 300-node subgraph of the 2002 Internet AS topology where filter deployment is above the critical density. We observe isolated pockets of size 1–4. Figure 29(b) shows pocket size distribution under a filter deployment below the critical filter density. We observe long distance linkages producing a large pocket. The critical filter density holds for CAIDA, RIPE, USC/ISI, and UMich topologies, and under different routing algorithms including semi-random and random routing.

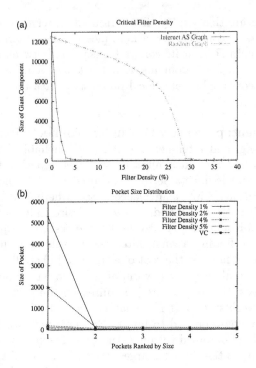

FIGURE 28   (a) Critical filter density for 12,514-node Internet AS graph vs. corresponding random graph. (b) Pocket size distribution under varying filter density ranked by size.

Finite time dynamics. Figure 30(a) shows worm propagation dynamics under random scanning for different filter densities. We observe that 2% filter deployment, even though ineffective asymptotically (cf. fig. 27), is able to slow the speed at which worms spread over finite time horizons. Figure 30(b) shows that the critical filter density is accordingly shifted to the left, whose magnitude depends on the finite time horizon. This may give a stripped down defense mechanism a little breathing room before it is fully activated, with the caveat that the deployed filters must already be able to detect the propagating worm. Thus the benefit of deploying fewer filters than mandated by the asymptotic critical filter density is largely restricted to repeat attacks of known worms where, for efficiency reasons, not all filter sites are primed against all known worms. Randomization over the set of known worm signatures coupled with on-demand filter loading may be needed depending on the overhead associated with mutation-resistant filtering and the size of the worm set.

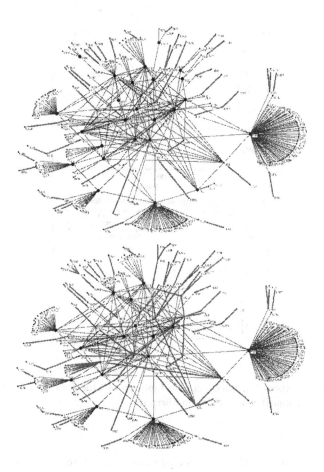

FIGURE 29   Worm propagation in Internet AS topology. Thick edges indicate pockets.
(a) Containment of infection. (b) Large-scale contamination.

Load-based filter placement. It turns out that a load-driven filter placement
strategy is able to achieve a 50% smaller asymptotic critical filter density than
VC-oriented filter placement. In one load-based filter placement algorithm, vertices are ordered by node load—actually a normalized variant that discounts
overlaps—and the highest ranked r nodes selected for inclusion in the filter net.
The r nodes are then removed from the graph to yield one or more connected
components. Node load is computed for the largest connected component, and
the selection and partitioning procedure recursively repeated until all connected
components are below a target size. Figure 31 shows the critical filter density
when $r = 5$. For very high degree nodes, the order dictated by node load is
nearly the same as that determined by degree; therefore their performance effect

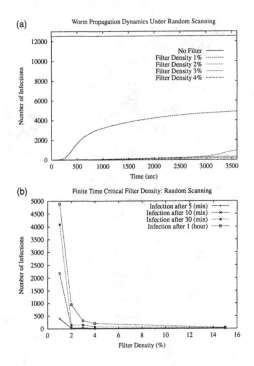

FIGURE 30   (a) Spread of infection as a function of time for different filter densities. (b) Critical filter density at finite time horizon.

is similar. However, for the intermediate load rank, filter selection deviates significantly between the two strategies, and node load more robustly identifies key junctions, the removal of which splits a single connected component into many smaller connected components.

4.5   DISCUSSION

The recently discovered power-law connectivity of various real-world networks, including Internet related graphs, helps address a long-standing question: What do real-world networks look like? Do they possess shared properties that may be used for performance evaluation? Although the area of power-law networks is still in its infancy, initial studies provide strong indications that power-law connectivity is a widespread phenomenon with repercussions to understanding, designing, and controlling networked systems. It is interesting that both network traffic and network topology owe much of their structure to heavy-tailed object sizes—files for self-similar traffic and neighborhood size in the case of power-law topology—that lead to self-similarity across multiple scales where a part resem-

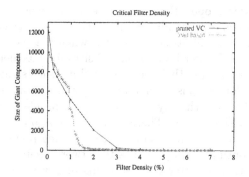

FIGURE 31 Critical filter density under load-based filter placement.

bles the whole. In the network connectivity context, this is also referred to as scale-freeness [12]: a random subgraph of a power-law graph inherits the parent's power-law degree distribution, a consequence of random sampling. For the same reason that self-similarity alone does not adequately characterize network traffic—for example, Brownian motion, when interpreted as network traffic, is self-similar but not long-range dependent—so does scale invariance only partially apply for network topology. A more essential property is power-law degree distribution, a trait distinct from self-similarity, that is responsible for generating a wide range of variability.

A modeling difference between self-similar traffic and power-law network topology lies in the explication of their causality. In self-similar network traffic, we can empirically ascertain that file sizes tend to be heavy-tailed, which provides the causal kernel upon which a reductionist explanation of the traffic phenomenon can rest. One may speculate why file sizes are heavy-tailed, but it is unclear that this leads to scientifically verifiable conclusions. For network topology, power-law connectivity appears to embody the dynamics principle "the rich get richer and the poor get poorer." Although appealing as a qualitative observation, as a quantitative explanation of why power-law connectivity arises in specific contexts, it misses several important ingredients. They include static design—airline and telecommunication networks, when initially laid out, are organized around hubs and POPs that are linked through a backbone, a design approach with a heavy dose of bootstrapped centralized management—and economic, social, and technological factors that can affect the growth and demise of individual components. A scientifically rigorous description of the origin of power-law connectivity seems to necessitate a better understanding of human behavior—collective and singular—a challenge transcending the confines of power-law connectivity.

Power-law networks open the door to fresh questions in optimization: are NP-hard graph problems in combinatorial optimization easier with respect to approximability, or even poly-time solvability, when graphs are power law? For example, the upper and lower bounds on VC sizes for Internet AS graphs show that VC approximation using the greedy algorithm yields significantly improved results vis-à-vis random graphs. Random walks, percolation, imperfect information games, and a host of other dynamic and static problems on graphs present new twists to established areas. There is, as yet, no good definition of deterministic power-law graphs, which is one of the starting points in this endeavor. The application of optimization problems to network security and resource provisioning indicates that a better understanding of power-law networks has the potential of approaching practical problems in novel ways.

Lastly, we remark that families of random graphs obeying a power-law degree distribution may, or may not, be sufficiently adequate to capture pertinent properties observed in real-world networks. Based on vertex cover size observations in real and artificial Internet AS graphs [103, 143], it appears that accurate capturing of vertex cover properties requires second-order connectivity structure, in addition to first-order structure in the form of power-law degree sequence. Refinements and their justification remain tasks for the future.

5   GAME THEORY

5.1   WHAT ARE NONCOOPERATIVE NETWORK GAMES?

A canonical example of a noncooperative network resource game that arises in the Internet context is congestion control. In a two-party congestion control game, Alice and Bob—two characters familiar in cryptography—share a common network resource that may get congested if too much traffic is submitted to the network. Congestion, technically, means that the total, or system-wide throughput declines if offered traffic exceeds a certain level. This is represented by a unimodal, dome-shaped load-throughput function [59, 102] (cf. fig. 32(a)) whose exact shape is dictated by the specific network context. One example, both old and new—old because the protocol in question has its roots in Abramson's ALOHA packet radio network [2] used in the early 1970s to connect the University of Hawaii's island campuses—is the throughput of wireless hot spots. In a wireless local area network (WLAN), resource sharing is governed by a competition-oriented protocol called carrier-sense multiple access with collision avoidance (CSMA/CA). A wireless station such as a laptop or handheld device listens to the radio channel (carrier sense), and if the channel is deemed idle, data transmission is attempted. If more than one station sends packets at about the same time (multiple access), the physical signals collide, leading to corruption. Collision is detected at the sender by the absence of an acknowledgment packet—in Ethernet, which uses a variant of the CSMA/CA protocol, collision

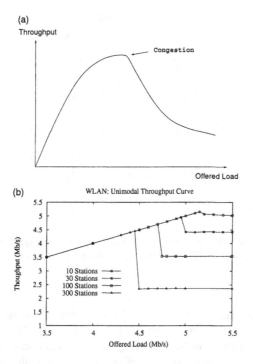

FIGURE 32   (a) Unimodal load-throughput curve and onset of congestion. (b) IEEE 802.11b WLAN throughput as a function of offered load for varying number of wireless stations.

can be detected at the physical layer without recourse to explicit acknowledgment (ACK) packets—at which time a form of exponential backoff in the retry time interval is instituted. That is, consecutive collisions lead to increasing pauses between retry attempts.

   All else being equal, the more stations and/or the higher the traffic demand at stations, the higher the probability of collision, which eventually leads to a decline in system throughput due to wasted bandwidth. Figure 32(b) shows the simulated load-throughput curve of IEEE 802.11b's CSMA/CA for 10, 30, 100, and 300 stations as a function of offered load. The sharp drop at a critical offered load stems from a correspondingly sharp increase in the collision rate. In real WLANs, throughput decline tends to be less pronounced due to biases resulting from uneven channel quality that skews the competition. In either case, there is a significant jump in the throughput variability within a session (over time) as well as across sessions at different wireless stations as the saturation point is crossed.

Bob

	C	N
C	5, 5	1, 9
N	9, 1	3, 3

Alice

FIGURE 33   Alice and Bob's throughput matrix for the congestion control game. C: cooperate, N: not cooperate. Throughput is measured in Mbps.

Returning to the congestion control game, it may be viewed as an instance of Prisoner's Dilemma mentioned in the Introduction, where Alice and Bob have recourse to two strategies upon encountering congestion: cooperation, which would mean backing off to help reduce total offered load, or selfishness, which may entail performing a congestion control action that does not reduce one's own traffic submitted to the network. Figure 33 shows a throughput matrix of Alice and Bob under all four combinations of congestion control strategies. When both parties cooperate (C), each achieves a throughput of 5 Mbps for a system throughput of 10 Mbps. When one is selfish but the other is not, the selfish party achieves a disproportionate share of 9 Mbps. When both are noncooperative, the system becomes congested with each receiving only 3 Mbps for a total throughput of 6 Mbps. In the throughput matrix example, Alice, if selfish and "rational," will choose the noncooperative strategy since her payoff—irrespective of Bob's action—always exceeds the corresponding throughput achievable by choosing the cooperative strategy: 9 Mbps vs. 5 Mbps when Bob chooses C, and 3 Mbps vs. 1 Mbps when Bob chooses N. Strategy N dominates strategy C. By symmetry, the game played by Alice and Bob results in the strategy profile (N,N) with payoff (3,3), which is strictly less than the "welfare" (5,5) attainable through cooperation.

Another example, in the context of TCP, is the throughput sharing behavior of multiple TCP sessions. Figure 34(a) shows the throughput achieved by five TCP sessions traversing a common bottleneck link. TCP, being cooperative, backs off when it thinks that a packet loss has occurred. This prevents congestion and promotes equitable sharing of resources when other extraneous factors such as distance—a long-haul TCP session is in a disadvantaged position vis-à-vis a

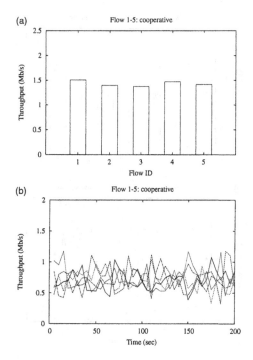

FIGURE 34   TCP dynamics: cooperative bandwidth sharing behavior. (a) Throughput share. (b) Time dynamics.

short-haul session—are ignored. Figure 34(b) shows the corresponding time dynamics of the bandwidth sharing behavior. Figure 35(a) shows the throughput attained by five TCP flows when one of them, Flow 5, implements a variant of TCP that incorporates a measure of selfishness. Upon detecting potential packet loss, "TCP-greedy" does not initially back off. Expecting other cooperative flows sharing the bottleneck to act gentlemanly, TCP-greedy persists at a larger time scale, soaking up the bandwidth abandoned by cooperative TCP flows. When throughput eventually degrades, TCP-greedy backs off, assuming that the losses are due to self-congestion. Figure 35(b) shows the time dynamics of a simple instance of TCP-greedy where, until time 100 second, the protocol operates cooperatively; at time 100 second, it toggles into greedy mode, which illustrates the contrast in bandwidth sharing behavior. We note that selfishness amid cooperation can be employed as a means of denial of service attack.

Congestion control games are single service class games where a shared resource is accessed through a single service class (that is, as is) representing the resource. The decision variable is: how much traffic to submit to the service

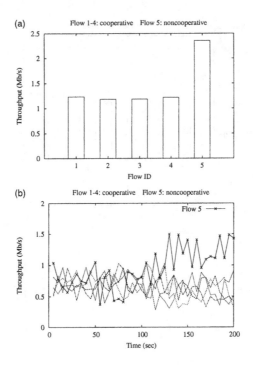

FIGURE 35 TCP dynamics: noncooperative bandwidth sharing behavior. (a) Throughput share. (b) Throughput dynamics—at time 100 second greedy action is instituted.

class or resource. In multiple service class games, a shared resource is accessed through a multiple service class abstraction, either because the underlying resource is actually a collection of distinct resources or a single resource is virtually divided into multiple resources through scheduling. The decision variables entail selecting service classes in addition to determining the traffic volume submitted. Multiclass network games are versatile—many network resource sharing problems can be cast as multiclass games—and possess a rich structure. They are discussed in section 5.3.

5.2 SINGLE-CLASS NONCOOPERATIVE NETWORK GAME: CONGESTION CONTROL

5.2.1 Equilibria and Optima.

Consider a binary congestion control game where Alice can send "excessive" offered load L_N or an "appropriate" traffic load L_C. With L_N and L_C replacing N and C, assume that the throughput matrix in figure 33 applies. In its continuous form (discussed in sec. 5.3) L_C and L_N are but two values of a traffic control variable $\lambda \in \mathbb{R}_+$. One of the first considerations, when analyzing dynamical systems—even the one-shot congestion control game—is the issue of stability. The configuration (N,N) with payoff (3,3) is an equilibrium in the sense that from (N,N) neither Alice nor Bob, acting selfishly and unilaterally, has an incentive to deviate since the payoff of the changing party would decrease to 1. Configurations conditioned on which unilateral strategy changes do not improve the changing party's individual utility are called Nash equilibria. In the two-party binary congestion control game—equivalent to Prisoner's Dilemma—strategy profile (N,N) is the unique Nash equilibrium. In general network resource games, including multi-class network games, a Nash equilibrium need not exist, much less be unique. Defining the system utility to be the sum of individual utilities, configurations (C,C), (C,N), and (N,C) are system optimal with total utility 10, whereas (N,N) with system utility 6 is not. A more refined, welfare-oriented notion of efficiency is Pareto optimality, where a configuration is Pareto optimal if the system utility cannot be improved without sacrificing the utility of one or more players. Thus (N,N) is not Pareto optimal since the configuration (C,C) improves both players' payoff; (C,C), (C,N), and (N,C) are Pareto optimal. It is easily checked that system optimality implies Pareto optimality; other implications, in general, do not hold. From a system and Pareto optimality perspective, configurations (C,C), (C,N), and (N,C) are equally good. Additional fairness criteria such as equal share may be imposed to further distinguish between desirable and undesirable system states.

5.2.2 **Pricing.** The binary congestion control game, being an instance of Prisoner's Dilemma, does not bring forth fundamental new issues when considered in the Internet context. However, it admits networking driven variations on the game with potential practical relevance. One case in point is pricing. By introducing usage pricing—the larger the bandwidth or data rate consumed, the higher the economic cost assigned by an ISP to the user—the dominance of noncooperative strategy N over cooperative strategy C may be removed. Supposing transferring a file at 1, 3, 5, and 9 Mbps costs a, b, c, and d dollars $(a < b < c < d)$, respectively, if Alice derives ordinal satisfaction

$$\langle 9 \text{ Mbps}, d \rangle \prec \langle 5 \text{ Mbps}, c \rangle \prec \langle 3 \text{ Mbps}, b \rangle \prec \langle 1 \text{ Mbps}, a \rangle$$

then strategy C becomes dominant over N. If Bob is equally cost-conscious and in no particular hurry, the solution to the pricing enhanced game with utility matrix shown in figure 36 becomes (C,C). In the case when Bob is a wealthy speed-junkie with utility preference

$$\langle 1 \text{ Mbps}, a \rangle \prec \langle 3 \text{ Mbps}, b \rangle \prec \langle 5 \text{ Mbps}, c \rangle \prec \langle 9 \text{ Mbps}, d \rangle$$

the solution of the game becomes (C,N) with throughput allocation (1,9). Bob is happy because he gets the fastest service, whereas Alice is happy because she gets the most economic service. The ISP is content because it only performs

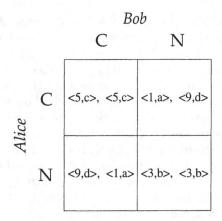

FIGURE 36 Alice and Bob's throughput-price matrix for the congestion control game with pricing $a < b < c < d$. C: cooperate, N: not cooperate.

metering and leaves the headache of resource contention resolution to users. The ISP would be happy if $a + d > 2c$. Pricing, when treated as an engineering tool, has the potential of steering away an otherwise self-defeating system from the peril of the tragedy of the commons [62]. Pricing, however, is not a magical wand and has limitations that can curtail its effectiveness. For example, pricing schemes that depend on accurate knowledge of users' utilities may be construed as unrealistic given the not so well understood nature of human preference. In some cases, pricing need not influence the outcome of a game. In the congestion control game, if both Alice and Bob are speed-junkies with unlimited spending power—e.g., $\langle 1\text{ Mbps}, a\rangle \prec \langle 3\text{ Mbps}, b\rangle$ and $\langle 5\text{ Mbps}, c\rangle \prec \langle 9\text{ Mbps}, d\rangle$ for all $a < b < c < d$—pricing does not prevent the congestion outcome (N,N).

5.2.3 Repeated Games.

Another direction in which the congestion control game can be looked at is as an iterated variant that injects a notion of time. Instead of playing the game in a single-shot fashion, a multiple round game is played where the players' future strategy is allowed to be influenced by the past. The Iterated Prisoner's Dilemma, highlighted in the influential work of Axelrod [7, 8], captures the problem of what time-dependent strategies are effective and, in an evolutionary twist, competitively fit when subject to evolutionary pressures. In tournaments where strategies were pitted against each other [7, 9], it was observed that one particular strategy, called tit-for-tat, outperformed the competition. In tit-for-tat, the strategy starts out cooperative and remains so as long as the other party reciprocates. If the other party switches to a noncooperative mode, tit-for-tat does the same. If the other party ever switches back to a cooperative mode, tit-for-tat is forgiving and reverts back as well. An unforgiving variant, called grim trigger, does not. In the finitely repeated Prisoner's Dilemma where the total payoff is the sum of the payoffs in each round, self-optimizing players will play the noncooperative strategy N in every round. If r is the final round, then no matter what has transpired in previous rounds, N is dominant over C in the final round by its finality—there are no future consequences for noncooperation. Applying the argument recursively to round $r - 1$ yields the all-N strategy. In an infinitely repeated Prisoner's Dilemma where the boundary condition of finite r is removed and future payoffs discounted by a multiplicative factor $0 < \delta < 1$, both tit-for-tat and grim trigger are self-optimizing strategies when δ is not too small. The two strategies induce an incentive for cooperation even when players are selfish. An excellent overview of repeated games can be found in Osborne and Rubinstein [99]. An introduction to evolutionary games is provided in Weibull [141].

In the Internet context, a single TCP-greedy session can exploit the cooperative nature of other TCP flows and grab a lion's share of network bandwidth. When many users employ greedy variants of TCP, the network may get trapped in a congested state. A survival of the fittest, "Wild Wild West" environment is not inconceivable. Technologically, deploying new transport protocols is not difficult, since it only entails changes on the sender and receiver ends. The network core, due to Internet's end-to-end design paradigm, is not affected. The outlook, however, is not singularly bleak. As indicated in the discussion on self-similar traffic, most TCP sessions are short-lived due to the heavy-tailed nature of file sizes. Short-lived sessions tend to operate in an aggressive mode called slow start—a misnomer given the fast nature of the start-up process—and rarely reach steady-state at which cooperativeness is most readily exposed. Also, a 100 msec small file transfer, even if tenfold delayed due to greedy actions of others, completes in one second which, on a human time scale, is not overly significant. A large file transfer with a 1-minute duration, if delayed tenfold, however, is a different matter. On the architectural front, users with broadband Internet connectivity are limited by an access speed of 1–6 Mbps which bounds the damage that a single user can exact on aggregate backbone traffic. For cellular data services, price-based controls, including usage pricing, are instituted to shield the low bandwidth cellular network from overutilization. Last, but not least, most TCP transfers are HTTP requests by clients to Web servers. As such, data flows downstream from server to client, and unless the server side TCP is modified, it is not easy to transform a TCP session into greedy mode with client side TCP modifications alone. An elegant treatment of congestion control from a distributed control and optimization perspective is provided in Kelly et al. [74].

5.3   MULTI-CLASS NONCOOPERATIVE NETWORK GAMES

5.3.1   Multiclass QoS Provisioning Game.

Consider a multiple service class resource provisioning system much like a multilane highway with m tollbooths that process arriving cars. The service quality—in the tollbooth example, delay—

received by an automobile depends on several factors including which toll lane the driver has joined and the line of cars waiting in front. All else being equal, the longer the line the longer the wait. In the Internet context, service may also be affected by a scheduler that apportions resources to the m classes according to a service policy. In the tollbooth example, suppose a single person is manning all booths, running from booth to booth collecting tolls. If toll lanes are differentiated by the number of passengers in a car including the driver—for example, lane 4 for cars with four or more passengers, lane 3 for three passengers, and so forth—and the tollbooth attendant implements a priority scheduling policy where cars in lane 3 are serviced only if lane 4 is empty, cars in lane 2 are serviced only if lanes 3 and 4 are empty, and so forth, then the quality of service received by a lone commuter may be less than that received by commuters in a carpool. In queuing theory, the equilibrium waiting time experienced by n customers in the m classes is studied as a function of the stochastic property of the customer arrival process and the scheduling policy implemented by the server [21]. In the network game context, we allow the n users to pick the service class to which their packets are submitted and focus on the outcome of the resultant n-player game. The impact of stochastic arrivals—a challenging problem for non-Markovian input in its own right—and time varying decision making by players is not explicitly considered except "in equilibrium" through static analysis.

Formally, an m-class n-player noncooperative network game comprises m service classes managed by a scheduler S and n players with traffic demand $\lambda_1, \lambda_2, \ldots, \lambda_n$ who choose one or more service classes via decision variables $\lambda_{ij} \geq 0$, $\lambda_i = \sum_{j=1}^{m} \lambda_{ij}$. A performance function φ—depending on S and the input—determines the quality of service $\varphi_1, \varphi_2, \ldots, \varphi_m$ rendered in the m service classes as a function of aggregate traffic $\mathbf{q} = (q_1, q_2, \ldots, q_m)$ where $q_j = \sum_{i=1}^{n} \lambda_{ij}$. The QoS received by the ith player is determined by his traffic allocation vector $\boldsymbol{\lambda}_i = (\lambda_{i1}, \lambda_{i2}, \ldots, \lambda_{im})$ and the choices made by other players. We assume $\partial\varphi_j/\partial q_j \geq 0$, that is, all else being equal, the more stress on a service class the worse the QoS—for example, delay or packet loss rate—rendered in that class. Assuming φ_j denotes packet loss rate, the throughput of service class j is given by $q_j(1 - \varphi_j)$. Pricing, if present, assigns a per unit flow cost p_j to each class. $\sum_j p_j \lambda_{ij}$ is the total cost incurred by user i and $\sum_j p_j q_j$ an ISP's total revenue. User i is endowed with a utility function U_i that depends on the traffic transmitted, QoS received, and cost. As a noncooperative game, user i's strategy set is given by the values of $\boldsymbol{\lambda}_i$, the system's strategy profile is $\boldsymbol{\lambda} = (\boldsymbol{\lambda}_1, \ldots, \boldsymbol{\lambda}_n)$, and user i aims to optimize his utility. A profit-maximizing service provider who sets the price vector $\mathbf{p} = (p_1, p_2, \ldots, p_m)$ may explicitly enter into play as player $n + 1$.

5.3.2 Applications.   A number of problems in networking can be mapped to the multiclass QoS provisioning game. A network access provider may furnish prioritized service classes—such as, platinum, gold, silver, bronze, and best-effort—that users can select depending on their application needs. For example, Alice, as an investment analyst, may use the gold class to browse the web and carry out e-commerce transactions at work, while using the bronze and best-effort classes at home for casual use. Bob, a financially strapped graduate student, may predominantly use best-effort service for Internet access except when videophoning Alice, at which time he uses platinum service. The INDEX project [5, 122] provides an interesting study of demand elasticity with respect to bandwidth pricing in an experimental multiclass access network. A network content provider or web service provider may use a multiclass service abstraction to schedule CPU cycles for exporting prioritized services to client requests. An enterprise network or transit provider may use multiclass packet forwarding to provide differentiated services to its customer base. Provisioning end-to-end QoS over a network of multiclass routers using game theoretic mechanisms is discussed in Chen and Park [27, 28]. In multipath routing, a set of routes—preferably disjoint—from source to destination is used to transmit traffic. In Orda et al. [98], game theoretic aspects of "parallel" routing are studied. By mapping each path in the parallel routing game to a service class, we arrive at an equivalent multiclass network game where the scheduling components $\varphi_1, \ldots, \varphi_m$ are mutually decoupled. Technically, the scheduler is a non-work conserving weighted fair queue (WFQ) with service weights $\alpha_1, \ldots, \alpha_m$, $\sum_j \alpha_j = 1$, that determine the bandwidth of each route/service class.

5.3.3 Key Issues and Analysis Tools.

Key issues when studying a multiclass QoS game include: Do multiclass QoS games possess Nash equilibria (NE)? If they do, are they efficient? Can pricing help drive the system to a desirable network state such as a system optimal NE? How are stability and optima affected by the shape of user utilities? How does scheduling impact the game? These are but a subset of questions with practical relevance.

Before we proceed with a discussion of game theoretic properties of multiclass QoS provisioning, we give a brief overview of some facts and analysis tools. First, in the following we consider only games with pure strategies, the default case implicitly assumed thus far. For example, in the binary congestion control game—a two-player finite game where finite refers to the cardinality of the strategy set $\{C, N\}$—mixed strategies would admit a probability distribution over $\{C, N\}$ with the interpretation that a player probabilistically chooses C or N. In the Internet QoS provisioning context, we do not believe that mixed strategies are practically meaningful. In so doing, however, we forgo one of the few niceties available in noncooperative games—the existence of a Nash equilibrium in finite noncooperative games with mixed strategies—a result established by Nash [94, 95]. Indeed, as we shall see, multiclass QoS games need not have a Nash equilibrium. This is not unusual since it is known that large finite games with random payoff functions, more often than not, do not possess an NE in pure strategies [43]. Mixed strategies are represented by simplices whose faces are spanned by corner points that correspond to pure strategies. As such, optimization tends to be easier under mixed strategies. For noncooperative games in pure strategies, a powerful result exists that assures the existence of a Nash equilibrium as long as the game is sufficiently nice in two respects: one, the strategy set of each player is nonempty, compact, and convex, and two, the utility function of every player i is continuous and quasi-concave in the ith decision variable λ_i. Recall that a function $f(x)$ is quasi-concave if for all a the upper level set $\{x : f(x) \geq a\}$ is convex. Satisfying the two conditions allows Kakutani's fixed point theorem—a generalization of Brouwer's fixed point theorem to point-set maps, that is, correspondences—to be applied yielding a fixed point to every user's self-optimizing action, called the best-reply correspondence. Recalling the definition of NE (cf. sec. 5.2), configuration $\lambda = (\lambda_1, \dots, \lambda_i, \dots, \lambda_n)$ is a Nash equilibrium if for all λ_i'

$$U_i(\lambda) \geq U_i(\lambda')$$

where $\lambda' = (\lambda_1, \dots, \lambda_i', \dots, \lambda_n)$. The maximizer λ_i', given a strategy profile λ—or best-reply correspondence—need not be unique leading to a point-set correspondence where Kakutani's fixed point theorem can come into play. In [121] it is shown that concave games with an additional requirement—diagonal strict concavity—possess a unique Nash equilibrium. The general sufficiency result for NE existence in concave games goes back to Debreu [40], Fan [51], and Glicksberg [60].

5.3.4 Stability and Efficiency.

We will first consider a multiclass analogue of Orda et al.'s parallel routing game [98] focusing on the role of utility functions, followed by the effect of pricing and scheduling in subsequent sections. As indicated earlier, the parallel routing game with m disjoint routes maps to an equivalent m-class QoS game where the scheduler S is trivial in the sense that the service classes are mutually decoupled. Let $\alpha = (\alpha_1, \dots, \alpha_m)$, $\sum_j \alpha_j = 1$, denote the relative bandwidth of the m service classes. A unit flow in class j receives resource share $\omega_j = \alpha_j / q_j$. We assume that QoS is determined by the per unit flow resource a traffic flow receives. That is, given λ and α, $\varphi_j \leq \varphi_{j'}$ if $\omega_j \geq \omega_{j'}$. A small value of φ means superior QoS. We consider utility functions motivated by certain multimedia applications. A VoIP application cannot tolerate delays more than 200 msec if it is to achieve a perceptually acceptable level of service. A real-time streaming video application cannot tolerate packet loss rates above a certain level, due to the negative repercussions on video quality stemming from undecodable video frames. A scenario that does not involve multimedia: a user running a supercomputing application needs to transfer a 125 GB file in less than three hours. This requires a bandwidth of at least 90 Mbps to achieve the QoS target. In all three cases, there is a sharp transition near a threshold such that QoS worse than the threshold is of little use and QoS better than the threshold brings marginal additional benefit: VoIP conversations sound clear whether packets are delayed 5 or 50 msec. Incorporating this property of QoS-sensitive applications, we consider step utility functions where the utility of user i has the

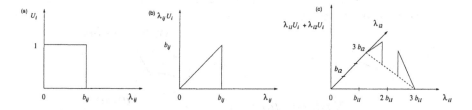

FIGURE 37   Utility function. (a) Single utility. (b) Weighted utility. (c) Combined utility.

shape

$$U_i(c) = \begin{cases} 1, & \text{if } c \leq \theta_i \,; \\ 0, & \text{otherwise}. \end{cases}$$

The expression $\theta_i \geq 0$ is a QoS threshold capturing i's preference and $c \geq 0$ is the QoS experienced. Assuming player i is allowed to split her total traffic λ_i among one or more classes—we also consider unsplittable games where all traffic belonging to user i must be sent to one class—we define user i's combined utility as $\bar{U}_i(\boldsymbol{\lambda}) = \sum_{j=1}^{m} \lambda_{ij} U_i(\varphi_j(\boldsymbol{\lambda}))$.

One more technical setup is needed before proceeding to a discussion of Nash equilibria. Since the service classes are decoupled and $\varphi_j(q_j)$ is nondecreasing in q_j, assuming φ_j is continuous we can express $U_i(c)$ directly in terms of a critical traffic threshold b_{ij} by the invertibility of φ_j:

$$U_i(\varphi_j(q_j)) = \begin{cases} 1, & \text{if } q_j \leq b_{ij} \,; \\ 0, & \text{otherwise}. \end{cases}$$

Figure 37 shows an example utility function for a two-class system where player i's QoS threshold θ_i maps to traffic thresholds b_{i1} and b_{i2} in classes 1 and 2, and total traffic demand is given by $\lambda_i = 3b_{i1} = 3b_{i2}$. Figure 37(a) shows utility function U_i for class $j \in \{1, 2\}$, Figure 37(b) shows the corresponding weighted utility $\lambda_{ij} U_i$, and Figure 37(c) shows the combined utility $\bar{U}_i$. Figure 37(c) shows that player i's utility is not quasi-concave in its decision variable $\boldsymbol{\lambda}_i$—the upper level set for small a has a hole in the middle—violating the sufficiency condition needed to apply Kakutani's fixed-point theorem. Not meeting the sufficiency condition for concave games does not imply that Nash equilibria do not exist. In Park et al. [109] it is shown that even in two-player two-service class systems, Nash equilibria fail to exist under "mild" conditions. Thus in these systems, every configuration has at least one player who thinks he can do better, preventing the system from reaching a quiescent state. The NE existence problem in the multiclass QoS game stems from two factors: multiple service classes—a distinguishing feature from the single-class congestion control game—and splitting of

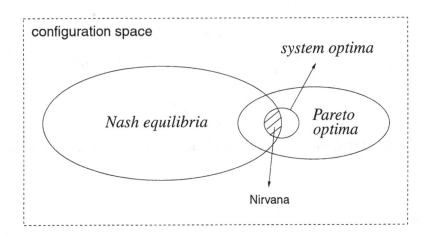

FIGURE 38 The structural relationship between three configuration classes—Nash equilibria, Pareto optima, and system optima—in multiclass QoS games.

user traffic. It can be shown that if traffic splitting is disallowed, the multiclass QoS game always has a Nash equilibrium.

The next question concerns the relationship between Nash equilibria, Pareto optima, and system optima. In particular, assuming NE exist, are they also efficient? The simple answer is "no." The three configuration classes—NE, Pareto optima, and system optima—for multiclass QoS games are well-separated, except for the trivial inclusion of system optima in Pareto optima. That is, there are NE that are not system optimal, an NE that is Pareto optimal need not be system optimal, a system optimum need not be a Nash equilibrium, and there are Pareto optima that are not system optimal. The structural relationship between the three classes is depicted in figure 38. In Park et al. [109] an exact characterization is given as to when Nash equilibria are Pareto or system optimal. A useful tool is the normal form of a configuration which establishes an equivalence relation between system optimality and Pareto optimality: a configuration λ is system optimal if and only if its normal form λ' is Pareto optimal. Through this linkage, system optimality of a configuration can be checked by verifying Pareto optimality of its normal form. Lastly, there are subclasses of multiclass QoS games, called resource-plentiful games, in which NE, Pareto optima, and system optima collapse into a single class.

5.3.5 A "Darker Side" of Pricing.

We illustrate that pricing in multiclass QoS games need not lead to desirable outcomes, in particular, we show that pricing can disrupt stability. In the parallel routing multiclass QoS game, consider pricing policies such that a traffic flow that receives better QoS—hence more resources per unit flow—pays a higher price than another that does not. Since service

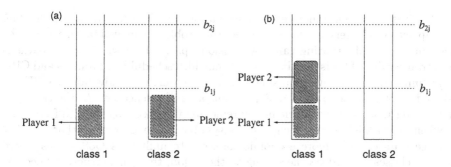

FIGURE 39 Detrimental effect of pricing in two-class two-player QoS game. (a) Nash equilibrium without pricing but unstable with pricing. (b) Configuration after player 2's selfish move; player 1 becomes unhappy.

classes are decoupled, the family of pricing functions that satisfy this property are monotone functions

$$\omega_j > \omega_{j'} \Leftrightarrow p_j > p_{j'}, \quad j \neq j'. \tag{7}$$

Such pricing functions are transparent, accurate, and fair—you pay for what you get. Now, consider a two-class two-player unsplittable QoS game where $b_{1j} < b_{2j}$ for $j \in \{1, 2\}$ (player 1 has a stricter QoS requirement than player 2), $\max\{\lambda_1, \lambda_2\} < \min\{b_{11}, b_{12}\}$ (both can be happy if there is one to each class), and $b_{1j} < \lambda_1 + \lambda_2 < b_{2j}$ for $j \in \{1, 2\}$ (if both are in the same class then player 1 will be unhappy while player 2 remains happy). Figure 39 shows an instance of the two-class two-player game. The configuration in figure 39(a) corresponds to a Nash equilibrium of the two-class two-player QoS game without pricing. Both players are happy and there is no incentive to move. When monotone pricing is introduced, however, the configuration ceases to be a Nash equilibrium since player 2 can move to service class 1 and still be happy QoS-wise while paying less money: $p_1 < p_2$ since $\omega_1 < \omega_2$. This is shown in figure 39(b). Player 1's QoS, however, is violated, which prompts player 1 to move to class 2. From here on the cycle repeats. This example illustrates that any usage-oriented—in the sense of eq. (7)—engineering application of pricing turns the QoS game with system optimal NE into one that is not even stable. Assuming utilities that dictate a user minimize cost subject to achieving QoS happiness, instability resulting from the "chasing a bargain and crowding" effect is inherent and can only be eliminated by violating eq. (7). The step function assumption on utilities can be relaxed without affecting the outcome of the game. The next section shows the beneficial effect of scheduling.

5.3.6 Influence of Scheduling.

In the example shown in figure 39, it is the excess capacity—hence cost—apportioned to player 2's flow when residing in class 2 that drives it to seek a more economic solution by migrating to class 1. As long as capacity in the service classes is fixed, the problem is unavoidable. Such is the case in the parallel routing game over disjoint paths: bandwidth across paths is not transferable. This is not the case in bandwidth scheduling at routers and CPU scheduling at servers, where a single physical resource—bandwidth or CPU—is shared among multiple flows under the auspices of a scheduler. In the two-class two-player QoS game with pricing, by removing the excess bandwidth from class 2 when player 2 is present, the gap $b_{22} - \lambda_2 > 0$ can be narrowed so that no other service class offers a more economic solution. In multiclass QoS games where service classes are ordered according to their QoS, the game structure becomes much nicer [119]. Nash equilibria are assured, and under "mild" conditions on utility, Nash equilibria become system optimal. The sufficiency condition on QoS ordering, called (A1), (A2), and (B) [119], is satisfied by the priority scheduler mentioned in the tollbooth example. Both QoS and usage-based price ordering across service classes is assured, and the resultant monotonicity helps break cyclic chain reactions that harm stability and efficiency.

A problem with priority scheduling is that lower priority classes may get starved when high priority traffic is abundant. WFQ, which assigns a fraction of the shared resource to classes in accordance with service weights, prevents starvation but does not satisfy properties (A1)–(B) needed for stable and efficient resource provisioning. It turns out that there is an optimal multiclass scheduler that satisfies the ordering properties and is maximally efficient in the mean-squared error (MSE) sense [119]. Intuitively, the optimal aggregate-flow scheduler is like a WFQ whose weights are dynamically set, that is, the weights are a function of the input. In Ren [116] an implementation of the optimal scheduler in IOS—Cisco's router operating system—and its benchmarking on a network of Cisco 7200 series routers is discussed. A corresponding simulation study is available in Ren and Park [118]. In Ren and Park [117] a queuing based treatment of optimal aggregate-flow scheduling is presented. Finally, we remark that when multidimensional QoS vectors are considered—for example, delay, packet loss rate, delay jitter, packet loss jitter, and bandwidth—a total order on the service classes is not guaranteed [29]. In general, only a partial order holds. For example, depending on the input and scheduling discipline, a high priority class with small delay or packet loss rate may experience a larger variance than a low priority class with larger delay or packet loss rate. Initial observations on multiclass QoS games with multidimensional QoS vectors may be found in Park et al. [109].

5.4 DISCUSSION

A game theoretic view of the Internet is interesting because a significant segment of its daily activity is carried out by network protocols, software that act on behalf of human users. The sometimes unpredictable human element is delimited by this layer of automated agents whose behavioral rules, for the most part, are transparent and remain invariant for months and years. Playing out game theory on the Internet provides a new opportunity to advance its theory as well as inject much needed effectiveness into a largely thought-experiment-driven, unverifiable subject matter. The scale of the Internet and its end-to-end design paradigm facilitate experimental design and quantitative studies where measurement based analysis of cause and effect can put the predictive power of theories to the test. This may complement the small scale experiments with human and animal subjects available today. Perhaps in a manner similar to self-similar Internet traffic and power-law Internet topology that have provided fresh jolts and experiential grounding to queuing theory and random graph theory, respectively, the Internet may provide a platform to further advance game theory. What is, as yet, missing is the identification of a phenomenological basis—long-range dependence in queuing and power-law connectivity in random graph theory—that directly connects with the foundations of game theory. Perhaps a quantification of the tragedy of the commons—very large or very small—and its linkage to "bounded rationality" and cooperativeness of present day protocols may be avenues for further exploration.

The discussion of network games presented is far from complete, omitting a number of influential works including market-based treatment of concave resource economies [52, 53, 123], noncooperative congestion control analyses [77, 78], pricing, flow control, and admission control in multiclass networks [34, 45, 65], among others. A more comprehensive overview of related works can be found in Park et al. [109]. We conclude by noting that two works stand out as milestones in the application of game theoretic ideas, including pricing, to network resource allocation. The Spawn system [139] by Huberman et al. at Xerox PARC in the early 1990s remains as one of the few efforts that have put microeconomic mechanisms to the test in a real working environment. The load balancing measurements enabled by Spawn have acted as a proof of concept for later works. In a similar vein, the INDEX project [122] by Varaiya et al. has served an important role by demonstrating the much-debated elasticity of humans—at least a 70+ population of students and faculty at Berkeley—with respect to access network bandwidth pricing.

6 CONCLUDING REMARKS

This chapter has attempted to provide a discussion of three subject areas—self-similar traffic, power-law connectivity, and noncooperative network games—whose underlying themes touch upon the "Internet as a Complex System" metaphor through synergistic but scientifically grounded efforts. The other two subject areas—scalable traffic control and organizational behavior—alluded to in the overall discussion, have been omitted due to time and space constraints. A complex systems viewpoint of the sciences, including natural and engineered phenomena, has proved both successful and unsuccessful. It has been successful because the ideas and challenges have helped the development of other sciences such as dynamical systems theory in mathematics. Also, many of the concepts and methods used in complex systems investigations—fractals, dynamics, randomness, automata, correlation at a distance, and phase transition, to mention a few—have entered into other areas, and, in some cases, into the everyday vocabulary of society at large. A complex systems viewpoint has been unsuccessful because the study of complex systems has not had a problem domain that it can claim as its own, at least not one that has yet withstood the test of time. Such a problem domain, perhaps, should be amenable to an arsenal of "complex systems tools" that are not necessarily in the purview of other sciences. This may not be a bad thing, but it is also a bit unsatisfactory because, to date, not many scientists would venture to say that they work in the area of complex systems, since it is unclear exactly what that means. It is too early to say if the Internet as a Complex System metaphor has depth and the requisite legs to go the distance. "System Science" is but one of the well-intentioned endeavors that has failed to put itself on terra firma. One of the positive aspects the Internet as a Complex System theme has going is that the subject matters discussed in the chapter were developed in other areas to solve domain-specific problems, but they also appear to require a unified understanding of the phenomena that transcend the scope of the individual areas. Some have started using the term "Internet Science" to refer to the broad menagerie of Internet related technologies, phenomena, and issues. We conjecture that a common fabric among the varied synergistic activities underlying the Internet may be a complex systems thread, meaning that in the Internet context, the whole is indeed greater than the sum of its parts, requiring new ways of bonding the varied elements from computer science, economics, mathematics, and social science.

7   ACKNOWLEDGMENTS

Supported in part by NSF grants ANI-9714707, ANI-9875789 (CAREER), ESS-9806741, EIA-9972883, ANI-0082861 (ITR), and grants from AFRL F30602-01-2-0539 (DARPA ATO FTN), Xerox, and ETRI.

REFERENCES

[1] Abello, J., A. Buchsbaum, and J. Westbrook. "A Functional Approach to External Graph Algorithms." In *Algorithms—ESA '98, 6th Annual European Symposium, Venice, Italy, August 24-26, 1998*, edited by G. Bilardi, G. F. Italiano, A. Pietracaprina, and G. Pucci, 332–343. Lecture Notes in Computer Science, vol. 1461. New York: Springer, 1998.

[2] Abramson, Norm. "The Aloha System—Another Alternative for Computer Communcations." In *Proc. Fall Joint Comput. Conf. AFIPS Conf.*, vol. 37, 281–285. Montvale, NJ: AFIPS Press, 1970.

[3] Aiello, W., F. Chung, and L. Lu. "A Random Graph Model for Massive Graphs." In *Proceedings of the 32nd Annual ACM Symposium on Theory of Computing; May 21–23, 2004, Portland, OR.* New York: ACM Press, 2000.

[4] Albert, R., H. Jeong, and A. Barabási. "Diameter of the World Wide Web." *Nature* **401** (1999): 130–131.

[5] Altmann, J., B. Rupp, and P. Varaiya. "Internet Demand under Different Pricing Schemes." In *Proceedings of the 1st ACM Conference on Electronic Commerce*, edited by S. Feldman and M. Wellman. New York: ACM Press, 1999.

[6] Asmussen, S., K. Binswanger, and B. Hojgaard. "Rare Events Simulation for Heavy-Tailed Distributions." *Bernouille* **6** (1991): 303–322.

[7] Axelrod, Robert. "Effective Choice in the Prisoner's Dilemma." *J. Conflict Res.* **24** (1980): 3–25.

[8] Axelrod, Robert. *The Evolution of Cooperation.* New York: Basic Books, 1984.

[9] Axelrod, R., and W. Hamilton. "The Evolution of Cooperation." *Science* **211** (1981): 1390–1396.

[10] Azar, Y., A. Fiat, A. Karlin, F. McSherry, and J. Saia. "Spectral Analysis of Data." In *Proceedings of the 33rd Annual ACM Symposium on Theory of Computing, July 6-8, 2001, Heraklion, Crete, Greece.* New York: ACM Press, 2001.

[11] Bailey, Norman. *The Mathematical Theory of Infectious Diseases.* New York: Oxford University Press, 1987.

[12] Barabási, A., and R. Albert. "Emergence of Scaling in Random Networks." *Science* **280** (1999): 509–512.

[13] Bellman, Richard. "On a Routing Problem." *Quart. Appl. Math.* **1** (1958): 87–90.

[14] Bennett, C., and D. DiVincenzo. "Quantum Information and Computation." *Nature* **404** (2000): 247–255.

[15] Bennett, C., and S. Wiesner. "Communication Via One- and Two-Particle Operators on Einstein-Podolsky-Rosen States." *Phys. Rev. Lett.* **69** (1992): 2881–2884.

[16] Beran, Jan. "A Test of Location for Data with Slowly Decaying Serial Correlations." *Biometrika* **76** (1989): 261–269.

[17] Beran, Jan. *Statistics for Long-Memory Processes.* Monographs on Statistics and Applied Probability. New York, NY: Chapman and Hall, 1994.

[18] Bollobás, B., O. Riordan, J. Spencer, and G. Tusnády. "The Degree Sequence of a Scale-Free Random Graph Process." *Random Struct. & Algorithms* **18(3)** (2001): 279–290.

[19] Boxma, O., and J. Cohen. "The Single Server Queue: Heavy Tails and Heavy Traffic." In *Self-Similar Network Traffic and Performance Evaluation*, edited by K. Park and W. Willinger, ch. 6. Wiley-Interscience, 2000.

[20] Braess, Dietrich. *Unternehmensforschung* **12** (1969): 258–268.

[21] Brandt, A., P. Franken, and B. Lisek. *Stationary Stochastic Models*. New York: John Wiley & Sons, 1990.

[22] Broder, A., R. Kumar, F. Maghoul, P. Raghavan, and R. Stata. "Graph Structure in the Web." *Comp. Networks* **33** (2000): 309–320.

[23] Cooperative Association for Internet Data Analysis (CAIDA). "Caida Analysis of Code-Red." 2001. ⟨http://www.caida.org/analysis/security/code-red⟩.

[24] Cooperative Association for Internet Data Analysis (CAIDA). "Skitter, a tool for actively probing the Internet in order to analyze topology and performance." 2002. ⟨http://www.caida.org/tools/measurement/skitter⟩.

[25] Computer Emergency Response Team (CERT). "CERT Advisory CA-1999-04 Melissa Macro Virus." March 1999. ⟨http://www.cert.org/advisories/CA-1999-04.html⟩.

[26] Computer Emergency Response Team (CERT). "CERT CERT Advisory CA-2001-19 'Code Red' Worm Exploiting Buffer Overflow In IIS Indexing Service DLL." July 2001. ⟨http://www.cert.org/advisories/CA-2001-19.html⟩.

[27] Chen, S., and K. Park. "An Architecture for Noncooperative QoS Provision in Many-Switch Systems." In *Proc. IEEE INFOCOM '99*, 864–872, 1999.

[28] Chen, S., and K. Park. "A Distributed Protocol for Multi-Class QoS Provision in Noncooperative Many-Switch Systems." In *Proceedings of the Sixth International Conference on Network Protocols*, 98–107. New York: IEEE Press, 1998.

[29] Chen, S., K. Park, and M. Sitharam. "On the Ordering Properties of GPS Routers for Multi-Class QoS Provision." In *Proc. SPIE International Conference on Performance and Control of Network Systems*, 252–265, 1998.

[30] Chuang, J., and M. Sirbu. "Pricing Multicast Communication: A Cost Based Approach." In *Proc. INET '98*, 1998. *langle*http://www.isoc.org/inet98/proceedings/6d/6d_2.htm⟩.

[31] Chung, F., and L. Lu. "The Average Distance in Random Graphs with Given Expected Degrees." *Proc. Natl. Acad. Sci.* **99** (2002): 15879–15882.

[32] Chung, F., and L. Lu. "Connected Components in a Random Graph with Given Degree Sequences." *Ann. Comb.* **6** (2002): 125–145.

[33] Clark, David. "The Design Philosophy of the DARPA Internet Protocols." In *Proc. ACM SIGCOMM '88*, vol. 18, no. 4 1988.

[34] Cocchi, R., D. Estrin, S. Shenker, and L. Zhang. "A Study of Priority Pricing in Multiple Service Class Networks." In *SIGCOMM '91, Proceedings of the Conference on Communications Architecture & Protocols, September 3-6, 1991, Zrich, Switzerland*. New York: ACM Press, 1991.

[35] Cox, D. R. "Long-Range Dependence: A Review." In *Statistics: An Appraisal*, edited by H. A. David and H. T. David, 55–74. Iowa State University Press, 1984.

[36] Crovella, M., and A. Bestavros. "Self-Similarity in World Wide Web Traffic: Evidence and Possible Causes." In *Proceedings of the 1996 ACM SIGMET-*

RICS *International Conference on Measurement and Modeling of Computer Systems*, 160–169. New York: ACM Press, 1996.

[37] Crovella, M., and L. Lipsky. "Long-Lasting Transient Conditions in Simulations with Heavy-Tailed Workloads." In *Proc. 1997 Winter Simulation Conference*, 1005–1012. New York: IEEE Press, 1997.

[38] Crovella, M., R. Frangioso, and M. Harchol-Balter. "Connection Scheduling in Web Servers." In *Proc. USENIX Symposium on Internet Technologies and Systems*. USENIX Assoc., 1999.

[39] Cunha, C., A. Bestavros, and M. Crovella. "Characteristics of WWW Client-Based Traces." Technical Report BU-CS-95-010, Computer Science Department, Boston University, 1995.

[40] Debreu, G. "A Social Equilibrium Existence Theorem." *Proc. Natl. Acad. Sci.* **38** (1952): 886–893.

[41] Devaney, Robert. *An Introduction to Chaotic Dynamical Systems.* Redwood City, CA: Addison-Wesley, 1985.

[42] Dijkstra, Edsger. "A Note on Two Problems in Connexion with Graphs." *Numerische Mathematik* (1959): 269–271.

[43] Dresher, M. "Probability of a Pure Equilibrium Point in n-Person Games." *J. Comb. Theory* **8** (1970): 134–145.

[44] Duffield, N. G., J. T. Lewis, N. O'Connel, R. Russell, and F. Toomey. "Statistical Issues Raised by the Bellcore Data." In *Proc. 11th IEE Teletraffic Symposium.* London: IEE, 1994.

[45] Dziong, Z., and L. Mason. "Fair-Efficient Call Admission Control Policies for Broadband Networks—A Game Theoretic Framework." *IEEE/ACM Trans. Network.* **4(1)** (1996): 123–136.

[46] Erdős, P., and T. Gallai. "Graphs with Points of Prescribed Degrees *(Hungarian)*." *Mat. Lapok* **11** (1960): 264–274.

[47] Erdős, P., and A. Rényi. "On Random Graphs." *Publ. Math. Debrecen* **6** (1959): 290–291.

[48] Erdős, P., and A. Rényi. "On the Evolution of Random Graphs." *Publ. Math. Inst. Hungar. Acad. Sci.* **5** (1960): 17–61.

[49] Erlang, A. K. "Solution of Some Problems in the Theory of Probabilities of Significance in Automatic Telephone Exchanges." *P. O. Elec. Eng. J.* **10** (1917): 189–197. Translated from the Danish 1917 article in *Elektroteknikeren*, vol. 13.

[50] Faloutsos, M., P. Faloutsos, and C. Faloutsos. "On Power-Law Relationships of the Internet Topology." In *Proc. ACM SIGCOMM '99*, 251–262, 1999. ⟨http://www.acm.org/sigs/sigcomm/sigcomm99/papers/session7-2.html⟩.

[51] Fan, K. "Fixed Point and Minimax Theorems in Locally Convex Topological Linear Spaces." *Proc. Natl. Acad. Sci.* **38** (1952): 121–126.

[52] Ferguson, D., C. Nikolaou, and Y. Yemini. "An Economy for Flow Control in Computer Networks." In *Proc. IEEE INFOCOM '89*, 110–118. Washington, DC: IEEE Press, 1989.

[53] Ferguson, D., Y. Yemini, and C. Nikolaou. "Microeconomic Algorithms for Load Balancing in Distributed Computer Systems." In *Proc. 8th International Conference on Distributed Computing Systems*, 491–499. New York: IEEE Press, 1988.

[54] Floyd, S., and V. Jacobson. "The Synchronization of Periodic Routing Messages." In *Proc. ACM SIGCOMM '93*, 33–44. New York: ACM Press, 1993.

[55] Ford, L., and D. Fulkerson. *Flows in Networks*. Princeton, NJ: Princeton University Press, 1962.

[56] Frieze, A., and C. McDiarmid. "Algorithmic Theory of Random Graphs." *Random Structures & Algorithms* **10** (1997): 5–42.

[57] Gallardo, J., D. Makrakis, and L. Orozco-Barbosa. "Fast Simulation of Broadband Telecommunications Networks Carrying Long-Range Dependent Bursty Traffic." *ACM Trans. Model. & Comp. Sim.* **11** (2001): 274–293.

[58] Garey, M., and D. Johnson. *Computers and Intractability*. San Francisco, CA: W. H. Freeman and Company, 1979.

[59] Gerla, M., and L. Kleinrock. "Flow Control: A Comparative Survey." *IEEE Trans. Commun.* **20(2)** (1980): 35–49.

[60] Glicksberg, I. L. "A Further Generalization of the Kakutani Fixed Point Theorem with Application to Nash Equilibrium Points." *Proc. Natl. Acad. Sci.* **38** (1952): 170–174.

[61] Harchol-Balter, M., and A. Downey. "Exploiting Process Lifetime Distributions for Dynamic Load Balancing." In *Proceedings of the 1996 ACM SIGMETRICS International Conference on Measurement and Modeling of Computer Systems*, 13–24. New York: ACM Press, 1996.

[62] Hardin, Garrett. "The Tragedy of the Commons." *Science* **162** (1968): 1243–1248.

[63] Heidelberger, Philip. "Fast Simulation of Rare Events in Queueing and Reliability Models." *ACM Trans. Model. & Comp. Sim.* **5** (1995): 43–85.

[64] Hethcote, Herbert. "The Mathematics of Infectious Diseases." *SIAM Rev.* **42(4)** (2000): 599–653.

[65] Hsiao, Man-Tung T., and Aurel A. Lazar. "Optimal Flow Control of Multi-Class Queueing Networks with Decentralized Information." In *Proc. IEEE INFOCOM '87*, 652–661. New York: IEEE Press, 1987.

[66] Hurst, Harold. "Long-Term Storage Capacity of Reservoirs." *Trans. Am. Soc. Civil Eng.* **116** (1951): 770–799.

[67] Husbands, P., H. Simon, and C. Ding. "On the Use of Singular Value Decomposition for Text Retrieval." In *Computation Information Retrieval*, edited by M. W. Berry. Proceedings in Applied Mathematics 106. SIAM, 2000.

[68] Irlam, Gordon. "Unix File Size Survey—1993." ⟨http://www.base.com/gordoni/ufs93.html⟩. September 1994.

[69] Jacobson, Van. "Congestion Avoidance and Control." In *Proc. ACM SIG-COMM '88*, 314–329. New York: ACM Press, 1988.

[70] Jain, R., and S. Routhier. "Packet Trains—Measurements and A New Model for Computer Network Traffic." *IEEE J. Select. Areas Commun.* **4(6)** (1986): 986–995.

[71] Jeong, H., B. Tomber, R. Albert, Z. Oltvai, and A. Babárasi. "The Large-Scale Organization of Metabolic Networks." *Nature* **407** (2000): 378–382.

[72] Jin, C., Q. Chen, and S. Jamin. "Inet: Internet Topology Generator." Technical Report CSE-TR-443-00, Department of EECS, University of Michigan, 2000.

[73] Kamath, K., H. Bassali, R. Hosamani, and L. Gao. "Policy-Aware Algorithms for Proxy Placement in the Internet." In *Proc. SPIE Scalability and Traffic Control in IP Networks*, vol. 4526, 157–171. SPIE, 2001.

[74] Kelly, F., A. Maulloo, and D. Tan. "Rate Control in Communication Networks: Shadow Prices, Proportional Fairness and Stability." *J. Oper. Res. Soc.* **49** (1998): 237–252.

[75] Kermack, W., and A. McKendrick. "A Contribution to the Mathematical Theory of Epidemics." *Proc. Roy. Soc. Lond. A* **115** (1927): 700–721.

[76] Kleinrock, Leonard. *Queueing Systems, Volume 1: Theory.* New York: Wiley-Interscience, 1975.

[77] Korilis, Y., and A. Lazar. "On the Existence of Equilibria in Noncooperative Optimal Flow Control." *J. ACM* **42(3)** (1995): 584–613.

[78] Korilis, Y., and A. Lazar. "Why is Flow Control Hard: Optimality, Fairness, Partial and Delayed Information." CTR Technical Report CU/CTR/TR 332-93-11, Center for Telecommunications Research, Columbia University, 1992. Presented in part in the Second ORSA Telecommunications Conference, 1992.

[79] Kumar, R., P. Raghavan, S. Rajagopalan, and A. Tomkins. "Trawling the Web for Emerging Cyber-Communities." *Comp. Net.* **31** (1999): 1481–1493.

[80] Labovitz, C., A. Ahuja, A. Bose, and F. Jahanian. "Delayed Internet Routing Convergence." In *Proc. of ACM SIGCOMM '00*, 175–187. New York: ACM Press, 2000.

[81] Leland, W., M. Taqqu, W. Willinger, and D. Wilson. "On the Self-Similar Nature of Ethernet Traffic." In *Proc. ACM SIGCOMM '93*, 183–193. New York: ACM Press, 1993.

[82] Liggett, Thomas. "Interacting Particle Systems." *Grundlehren der Mathematischen Wissenschaften*, vol. 276. New York: Springer Verlag, 1985.

[83] Liu, Z., P. Nain, D. Towsley, and Z. Zhang. "Asymptotic Behavior of a Multiplexer Fed by a Long-Range Dependent Process." *J. Appl. Prob.* **36(1)** (1999): 105–118.

[84] Lu, Linyuan. *Probabilistic Methods in Massive Graphs and Internet Computing.* Ph.D. thesis, University of California, San Diego, 2002.

[85] Makowski, A., and M. Parulekar. "Buffer Asymptotics for $M/G/\infty$ nput Processes." In *Self-Similar Network Traffic and Performance Evaluation*, edited by K. Park and W. Willinger, ch. 10. Wiley-Interscience, 2000.

[86] Mandelbrot, Benoit. "A Note on a Class of Skew Distribution Function: Analysis and Critique of a Paper by H. A. Simon." *Infor. & Control* **2** (1959): 90–99.

[87] Mandelbrot, B., and J. Van Ness. "Fractional Brownian Motions, Fractional Noises and Applications." *SIAM Rev.* **10** (1968): 422–437.

[88] McCorkendale, B., and P. Ször. "Code Red Buffer Overflow." *Virus Bulletin*, 4–5, September 2001.

[89] Mercator Internet AS Map. Courtesy of Ramesh Govindan, USC/ISI, 2002.

[90] Mikosch, T. "Regular Variation, Subexponentiality and Their Applications in Probability Theory." Technical Report, Eurandom, 1999.

[91] Milgram, Stanley. "The Small World Problem." *Psychol. Today* **2** (1967): 60–67.

[92] Molloy, M., and B. Reed. "The Size of the Giant Component of a Random Graph with a Given Degree Sequence." *Combin. Probab. Comput.* **8** (1998): 295–305.

[93] Moore, D., G. Voelker, and S. Savage. "Inferring Internet Denial-of-Service Activity." In *Proc. USENIX Security Symposium*, 9–22. USENIX Assoc., 2001.

[94] Nash, J. F. "Equilibrium Points in n-Person Games." *Proc. Natl. Acad. Sci.* **36** (1950): 48–49.

[95] Nash, J. F. "Non-Cooperative Games." *Ann. Math.* **54** (1951): 286–295.

[96] National Laboratory for Applied Network Research. Routing data, 2000. ⟨http://moat.nlanr.net/Routing/rawdata/⟩.

[97] Newman, M. "The Structure of Scientific Collaboration Networks." *Proc. Natl. Acad. Sci.* **4** (2001): 404–409.

[98] Orda, A., R. Rom, and N. Shimkin. "Competitive Routing in Multiuser Communication Networks." *IEEE/ACM Trans. Net.* **1(5)** (1993): 510–521.

[99] Osborne, M., and A. Rubinstein. *A Course in Game Theory.* The MIT Press, 1994.

[100] Östring, S., H. Sirisena, and I. Hudson. "Rate Control of Elastic Connections Competing with Long-Range Dependent Network Traffic." *IEEE Trans. Commun.* 2001.

[101] Page, L., S. Brin, R. Motwani, and T. Winograd. "The Pagerank Citation Ranking: Bringing Order to the Web." Stanford Digital Libraries Technologies Project, Stanford, California, 1998. ⟨http://dbpubs.stanford.edu/pub/1999-66⟩.

[102] Park, Kihong. "Warp Control: A Dynamically Stable Congestion Protocol and Its Analysis." In *Proc. ACM SIGCOMM '93*, 137–147, 1993.

[103] Park, K., and H. Lee. "On the Effectiveness of Route-Based Packet Filtering for Distributed DoS Attack Prevention in Power-Law Internets." In *Proceedings of the 2001 SIGCOMM Conference on Applications, Technolo-*

gies, Architectures, and Protocols for Computer Communications, 15–26. New York: ACM Press.

[104] Park, K., and T. Tuan. "Performance Evaluation of Multiple Time Scale TCP under Self-Similar Traffic Conditions." *ACM Trans. Model. & Comp. Sim.* **10(2)** (2000): 152–177.

[105] Park, K., and W. Wang. "QoS-Sensitive Transport of Real-Time MPEG Video Using Adaptive Forward Error Correction." In *Conference on Multimedia Computing and Systems*, vol. II, 426–432. Washington, DC: IEEE Press, 1999.

[106] Park, K., and W. Wang. "QoS-Sensitive Transport of Real-Time MPEG Video Using Adaptive Redundancy Control." *Comp. Comm.* **24** (2001): 78–92.

[107] Park, K., and W. Willinger. "Self-Similar Network Traffic: An Overview." In *Self-Similar Network Traffic and Performance Evaluation*, edited by K. Park and W. Willinger. Wiley-Interscience, 2000.

[108] Park, K., G. Kim, and M. Crovella. "On the Relationship between File Sizes, Transport Protocols, and Self-Similar Network Traffic." In *Proc. IEEE International Conference on Network Protocols*, 171–180, October 1996. ⟨http://citeseer.ist.psu.edu/park96relationship.html⟩.

[109] Park, K., M. Sitharam, and S. Chen. "Quality of Service Provision in Noncooperative Networks: Heterogeneous Preferences, Multi-Dimensional QoS Vectors, and Burstiness." In *Proc. 1st International Conference on Information and Computation Economies* (ICE-98), 111–127. New York: ACM Press, 1998.

[110] Paxson, V., and S. Floyd. "Wide-Area Traffic: The Failure of Poisson Modeling." *IEEE/ACM Trans. on Networking* **3(3)** (1995): 226–244.

[111] Phillips, G., S. Shenker, and H. Tangmunarunkit. "Scaling of Multicast Trees: Comments on the Chuang-Sirbu Scaling Law." In *Proc. ACM SIGCOMM '99*, 41–51. New York: ACM Press, 1999.

[112] Pikovsky, A., M. Rosenblum, and J. Kurths. *Synchronization*. Cambridge, UK: Cambridge University Press, 2001.

[113] Rabin, Michael. "Efficient Dispersal of Information for Security, Load Balancing, and Fault Tolerance." *J. ACM* **36(2)** (1989): 335–348.

[114] Redner, S. "How Popular is Your Paper?" *Eur. Phys. J. B* **4** (1998): 131–134.

[115] Reed, I., and G. Solomon. "Polynomial Codes Over Certain Finite Fields." *J. SIAM* **(10)** (1960): 300–304.

[116] Ren, Huan. *Aggregate-Flow Scheduling: Theory and Practice*. Ph.D. thesis, Purdue University, 2002.

[117] Ren, H., and K. Park. "On the Complexity of Optimal Aggregate-Flow Scheduling." Technical Report CSD-TR-02-021, Department of Computer Sciences, Purdue University, 2002.

[118] Ren, H., and K. Park. "Performance Evaluation of Optimal Aggregate-Flow Scheduling: A Simulation Study." *Comp. Comm.* (2002): to appear.

[119] Ren, H., and K. Park. "Toward a Theory of Differentiated Services." In *Proc. IEEE/IFIP Eighth International Workshop on Quality of Service*, 211–220. Washington, DC: IEEE Press, 2000.

[120] RIPE NCC. "Routing Information Service Raw Data." 2002. ⟨http://data.ris.ripe.net⟩.

[121] Rosen, J. B. "Existence and Uniqueness of Equilibrium Points for Concave n-Person Games." *Econometrica* **33(3)** (1965): 520–534.

[122] Rupp, B., R. Edell, H. Chand, and P. Varaiya. "INDEX: A Platform for Determining how People Value the Quality of Their Internet Access." In *Proc. IEEE/IFIP IWQoS*, 85–90. Washington, DC: IEEE Press, 1998.

[123] Sairamesh, J., D. Ferguson, and Y. Yemini. "An Approach to Pricing, Optimal Allocation and Quality of Service Provisioning in High-Speed Networks." In *Proc. IEEE INFOCOM '95*, 1111–1119. Washington, DC: IEEE Press, 1995.

[124] Shaikh, A., J. Rexford, and K. Shin. Load-sensitive routing of long-lived IP flows. In *Proc. ACM SIGCOMM '99*, 1999. ⟨http://citeseer.ist.psu.edu/article/shaikh99loadsensitive.html⟩.

[125] Shannon, C., and W. Weaver. *The Mathematical Theory of Communication*. Urbana, IL: The University of Illinois Press, 1949.

[126] Simon, H. A. "On a Class of Skew Distribution Functions." *Biometrica* **42** (1955): 425–440.

[127] Spafford, Eugene. "The Internet Worm: Crisis and Aftermath." *Comm. ACM* **32(6)** (1989): 678–687.

[128] Szpankowski, W., P. Jacquet, and L. Georgiadis. Personal communication, 2001.

[129] Tangmunarunkit, H., J. Doyle, R. Govindan, S. Jamin, W. Willinger, and S. Shenker. "Does AS Size Determine AS Degree?" *Comp. Comm. Rev.* **31(5)** (2001).

[130] Tangmunarunkit, H., R. Govindan, and S. Shenker. "Internet Path Inflation Due to Policy Routing." In *Proc. SPIE Scalability and Traffic Control in IP Networks*, 188–195. CAIDA, 2001. ⟨http://www.caida.org/outreach/papers/2001/OSD/⟩.

[131] Taqqu, M., W. Willinger, and R. Sherman. "Proof of a Fundamental Result in Self-Similar Traffic Modeling." *Comp. Comm. Rev.* **26** (1997): 5–23.

[132] Tuan, T., and K. Park. "Multiple Time Scale Congestion Control for Self-Similar Network Traffic." *Perf. Eval.* **36** (1999): 359–386.

[133] Tuan, T., and K. Park. "Multiple Time Scale Redundancy Control for QoS-Sensitive Transport of Real-Time Traffic." In *Proceedings of Infocom '00: Nineteenth Annual Joint Conference of the IEEE Computer and Communications Societies*, 1683–1692. New York: IEEE Press, 2000.

[134] University of Michigan. AS Graph Data Sets, 2002. ⟨http://topology.eecs.umich.edu/data.html⟩.

[135] University of Oregon. Oregon Route Views, 2000. ⟨http://www.routeviews.org/ and http://archive.routeviews.org/⟩.

[136] Veres, A., and M. Boda. "The Chaotic Nature of TCP Congestion Control."
In *Proc. IEEE INFOCOM '00*, 1715–1723, 2000.

[137] Veres, A., Z. Venesi, S. Molnar, and G. Vattay. "On the Propagation
of Long-Range Dependence in the Internet." In *Proceedings of Infocom '00: Nineteenth Annual Joint Conference of the IEEE Computer
and Communications Societies*, 243–254. New York: IEEE Press, 2000.
⟨http://citeseer.ist.psu.edu/article/veres00propagation.html⟩.

[138] von Neumann, J., and O. Morgenstern. *Theory of Games and Economic
Behavior.* Princeton, NJ: Princeton University Press, 1944.

[139] Waldspurger, C., T. Hogg, B. Huberman, J. Kephart, and W. Stornetta.
"Spawn: A Distributed Computational Economy." *IEEE Trans. Software
Eng.* **18(2)** (1992): 103–117.

[140] Watts, D., and S. Strogatz. "Collective Dynamics of 'Small World' Networks." *Nature* **393** (1998): 440–442.

[141] Weibull, Jörgen. *Evolutionary Game Theory.* Cambridge, MA: MIT Press,
1997.

[142] Willinger, W., M. Taqqu, R. Sherman, and D. Wilson. "Self-Similarity
through High-Variability: Statistical Analysis of Ethernet LAN Traffic at
the Source Level." In *Proc. ACM SIGCOMM '95*, 100–113. New York:
ACM Press, 1995.

[143] Winick, J., and S. Jamin. "Inet-3.0: Internet Topology Generator." Technical Report CSE-TR-456-02, Department of EECS, University of Michigan,
2002.

[144] Wu, G., E. Chong, and R. Givan. "Congestion Control via Online Sampling." In *Proceedings of Infocom '01: Twentieth Annual Joint Conference
of the IEEE Computer and Communications Societies*, 1271–1280. New
York: IEEE Press, 2001.

[145] Zhang, L., and D. Clark. "Oscillation Behavior of Network Traffic: A Case
Study Simulation." *Internetworking: Res. & Exper.* **1(2)** (1990): 101–112.

[146] Zipf, G. *Human Behavior and the Principle of Least Effort.* Cambridge,
MA: Addison-Wesley Press, 1949.

Passive Traffic Measurement for Internet Protocol Operations

Matthias Grossglauser
Jennifer Rexford

1 INTRODUCTION

Managing a large communication network is a complex task handled by a group of human operators. These operators track the state of the network to detect equipment failures and shifts in traffic load. After detecting and troubleshooting a problem, they may change the configuration of the equipment to improve the utilization of network resources and the performance experienced by end users. The ability to detect, diagnose, and fix problems depends on the information available from the underlying network. These tasks are easier if the equipment provides reliable and predictable service and retains detailed information about the resources consumed by the ongoing data transfers. In diagnosing potential performance problems, operators benefit from having an accurate and timely view of the flow of traffic across the network. As part of fixing a problem, operators need to know in advance how changes in configuration might affect the distribution of traffic and the performance experienced by end users.

The Internet as a Large-Scale Complex System,
edited by Kihong Park and Walter Willinger, Oxford University Press.

The goals of network operators are in direct conflict with the design of the Internet protocol (IP). The primary goal in the design of the ARPAnet in the late 1960s was to make efficient use of existing data networks through the use of packet switching [8]. End hosts divide data into packets that flow through the network independently. In forwarding packets toward their destinations, the network routers do not retain information about ongoing transfers and do not provide fine-grained support for performance guarantees. As a result, packets may be corrupted, lost, delayed, or delivered out of order. This complicates the efforts of network operators to provide predictable communication performance for their customers. Rather than having complexity inside the network, the end hosts have the responsibility for the reliable, ordered delivery of data between applications. Implemented on end hosts, the transmission control protocol (TCP) plays a crucial role in providing these services and adapting to network congestion. Inside the network, the routers implement routing protocols that adapt to equipment failures by computing new paths for forwarding IP packets. These automatic and distributed reactions to congestion and failures make it difficult for network operators to detect, diagnose, and fix potential problems.

The commercial realities of today's Internet introduce additional challenges for network operators. The Internet consists of tens of thousands autonomous systems (ASes), where each AS is a collection of routers and links managed by an single institution, such as a company, university, or Internet service provider (ISP). Within an AS, the routers communicate via an intradomain routing protocol such as open shortest path first (OSPF) or intermediate system-intermediate system (IS-IS). The operators configure each unidirectional link with an integer weight, and the routers distribute these weights and use them to compute shortest paths through the network, as shown in figure 1. Neighboring ASes use the border gateway protocol (BGP) to exchange information about how to reach destinations throughout the Internet, without sharing the details of their network topologies and routing policies. Much of the traffic flowing through the Internet must traverse multiple ASes. As such, the operators of any individual AS do not have end-to-end control of the distribution of traffic. The lack of end-to-end control makes it difficult for operators to detect and diagnose performance problems, and predict the impact of changes in network configuration.

Traffic measurement plays a crucial role in providing operators with a detailed view of the state of their networks. Operators detect congestion based on periodic summaries of traffic load and packet loss on individual links. By sending active probes between pairs of points in the network, operators can identify parts of the network that exhibit high delay or loss, or detect routing anomalies such as forwarding loops. Measurement also helps identify which traffic is responsible for network congestion, and whether the additional packets stem from a denial-of-service attack, a popular event or service on the web, or a routing change caused by an equipment failure or a reconfiguration in a different AS. Finally, measurement can assist operators in evaluating possible changes to the network configuration. Operators may alleviate congestion by changing the parameters

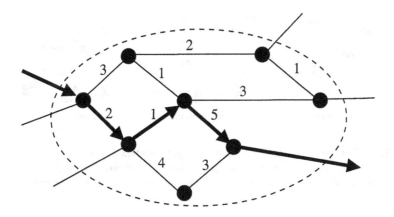

FIGURE 1   Shortest path routing within a domain based on OSPF/IS-IS link weights.

that control the allocation of resources in the network. For example, changing one or more OSPF/IS-IS weights would trigger the selection of new paths, which could result in a more efficient flow of traffic through the network. Operators may also change the configuration of the buffer management or link scheduling parameters, or install a packet filter to discard malicious traffic.

Diagnosing performance problems and evaluating the effectiveness of corrective actions depends on having accurate information about the flow of traffic through the network. In section 2, we present three models that provide a network-wide view of the traffic in an AS. We define the path, traffic, and demand matrices and describe how operators can use these models to diagnose and fix performance problems in their networks. Populating these models requires collecting extensive traffic measurements from multiple locations in an operational network. However, most IP networking equipment does not include the necessary support for collecting this data. In section 3 we survey existing techniques for collecting traffic statistics in IP networks. We present a brief overview of SNMP/RMON, packet monitoring, and flow-level measurement. Constructing a network-wide view of the traffic requires new measurement techniques or the careful combination of several types of measurement data from multiple locations in the network. In section 4, we describe recent research work on computing path, traffic, and demand matrices. We conclude the chapter in section 5 with a summary and a discussion of future research directions.

2   NETWORK-WIDE TRAFFIC MODELS

Network-wide representations of the traffic are necessary to drive network-wide control actions, such as routing changes, traffic classification and filtering, and

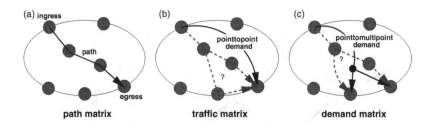

FIGURE 2 Network-wide traffic models: (a) path matrix (volume for every path be-tween every ingress-egress pair); (b) traffic matrix (volume for every ingress-egress pair); (c) demand matrix (volume from every ingress to every set of egress points).

capacity planning. In this section, we introduce three canonical spatial repre-sentations of the traffic—the *path matrix*, the *traffic matrix*, and the *demand matrix*—and discuss the application of these models in managing an IP net-work. These representations cover a spectrum from *descriptive* (i.e., capturing precisely the current state of the network and the current flow of traffic), to *predictive* (i.e., enabling the prediction of the state of the network after taking hypothetical control actions).

2.1   PATH, TRAFFIC, AND DEMAND MATRICES

To motivate these three canonical representations, consider how a network op-erator would perform traffic engineering if he had complete, end-to-end control over the Internet. There are two natural representations of the traffic that would help the operator detect and diagnose problems and evaluate potential control actions. The first representation is the *path matrix*, which specifies a data vol-ume $V(P)$ over some time interval of interest for every path P between every source host S and destination host D,[1] as shown in figure 2(a). The path matrix captures the current state and behavior of the network by providing a detailed description of the spatial flow of traffic through the network.

The second representation is the *traffic matrix*, which specifies a data vol-ume $V(S, D)$ per source-destination pair (S, D), as shown in figure 2(b). The traffic matrix captures the *offered load* on the network. This representation is based on the idealized assumption that the offered load is *invariant* to the state and behavior of the network. As such, in combination with a network model, the traffic matrix allows the operator to predict the flow of traffic through the network after hypothetical changes have been made. The operator can decide on a control action that will have the desired impact on the traffic flow.

---

[1]Strictly speaking, given that packets can be dropped at routers inside the network, the path matrix should also include every path from a source S to an internal node (router) R.

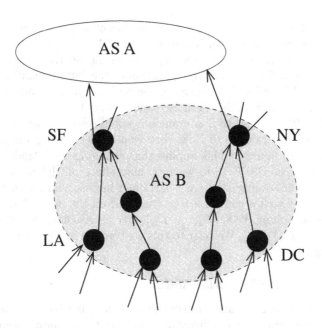

FIGURE 3 Forwarding traffic to the closest egress point.

In practice, a network operator only has control over a small portion, or autonomous system (AS), of the Internet. Capturing the traffic in a single AS suggests a slightly richer set of models. The first two representations are directly inspired by the path and traffic matrix, with the caveat that the path matrix only covers the set of paths within the AS and the traffic matrix is expressed between ingress-egress nodes rather than source-destination hosts. These two representations would suffice in a situation where interdomain and intradomain routing were completely decoupled, that is, if local control actions could have no impact on where traffic enters and leaves the AS. Under this assumption, managing a single AS is conceptually similar to managing the Internet end-to-end, albeit at a smaller scale. However, in reality, interdomain and intradomain routing interact with each other. Routing between ASes depends on the border gateway protocol (BGP) [36], a path-vector protocol that computes routes at the AS level for blocks of destination IP addresses. BGP computes a sink tree spanning ASes all over the Internet for each prefix. Each block (or, *prefix*) consists of a 32-bit address and a mask length (e.g., 192.0.2.0/24 refers to the IP addresses ranging from 192.0.2.0 to 192.0.2.255).

An AS learns routes to various destination prefixes from neighboring domains. Each router in the AS selects a *best* path for each destination prefix based on the BGP advertisements and the local routing policies. The crucial point is that the BGP routing process may select a *set* of egress points for a given prefix, rather than a single egress point [11, 13]. The selection of a single egress point at any given router depends on the relative proximity to these egress points, as defined by the configuration of the *intra*domain routing protocol, such as OSPF or IS-IS. For example, suppose that AS A advertises a route to 192.0.2.0/24 to AS B in San Francisco and in New York. Then, routers "close" (in the OSPF/IS-IS sense) to San Francisco would select the egress point in San Francisco, and routers close to New York would select the egress point in New York, as shown in figure 3. This implies that changing the intradomain routing parameters, such as OSPF/IS-IS weights, may affect which egress point is used to reach the destination prefix from a particular ingress point. To accurately predict the effect of such a control action, the load on the network domain has to be expressed as volume $V(S, \{D_1, \ldots, D_n\})$ per ingress node S and *set* of egress nodes $\{D_1, \ldots, D_n\})$ [14]. We refer to this abstraction as the *demand matrix*, as shown in figure 2(c).[2]

2.2   NETWORK OPERATIONS TASKS

The path, traffic, and demand matrices provide a sound basis for detecting and diagnosing problems, and for "what-if" analysis to evaluate potential control actions. This is important for a variety of important tasks in managing an IP network.

Generating traffic reports for customers: A service provider typically generates basic traffic reports for customers, often as part of billing for access to the network. A large customer may connect to the provider via multiple access links in different parts of the country. The provider can use the traffic matrix to generate reports of the volume of traffic between each pair of access links for these customers. The path matrix can be used to demonstrate how traffic to/from the customer flows through the provider's backbone by focusing on the subset of paths that start or end at the customer's access link(s). The path matrix also allows the provider to determine if the customer's traffic traverses any congested links. This analysis allows the provider to identify the customers that may be experiencing bad performance during a congestion event, or to demonstrate that a customer's performance problem arises in some other part of the end-to-end path, outside of the provider's network.

Diagnosing the cause of congestion: Network operators typically monitor aggregate statistics about traffic volume and packet loss on a per-link basis, as discussed later in section 3.1. Although these measurements are sufficient to identify overloaded links, diagnosing the cause of the congestion requires fine-grained information about the flow of traffic into and out of the link. The path matrix allows the operator to determine the set of paths and their associated traffic

---

[2] Alternatively, volume could be expressed as point to point, but over an extended topology that adds virtual, zero-weight, output links from the set of possible egress points to a virtual end node representing the set of destination prefixes sharing the same egress set.

intensities (and possibly other attributes). For example, a distributed denial-of-service (DDoS) attack would cause patterns that resemble a sink tree, with a large amount of traffic entering at multiple ingress points, and leaving at only one or a small number of egress points, probably toward a single destination address. On the other hand, congestion caused by a forwarding loop would manifest itself through the presence of cyclic paths that contain the overloaded link.

Tuning intradomain routing to alleviate congestion: Suppose that the network operators have identified a set of ingress-egress pairs responsible for congestion on one or more links in the backbone. The operators could examine the evolution of the traffic matrix over time to assess whether the heavy load for these ingress-egress pairs is a persistent (and perhaps worsening) problem and whether it arises at certain times of the day. The operators can fix the problem by reconfiguring the OSPF/IS-IS weights or MPLS (multi-protocol label switching) label-switched paths to move some traffic off of the congested links. The traffic or demand matrix is a crucial part of evaluating potential configuration changes. For a proposed set of link weights W, say, the demand matrix allows the operator to predict the load on each link l as $\sum_{(S,D):l \in p(S,D,W)} V(S,D)$, where $p(S,D,W)$ is the shortest path between S and D under link weights W. The traffic matrix is sufficient if the routing change, say the rerouting of an MPLS label-switched path, would not affect how traffic exits the network.

Planning changes to the network topology: At a longer time scale (weeks to months), the path, traffic, and demand matrices can drive capacity planning decisions. Studying the evolution of the path matrix over time can help identify potential bottleneck links that should be upgraded to higher capacity. The traffic and demand matrices over time can be used to evaluate how the deployment of new routers and links would affect the flow of traffic in the network. The demand matrix can be used to evaluate decisions about whether and where to add another link (i.e., new egress points) to an existing peer or customer. In addition, these predictive models can be used to evaluate the impact of adding a new customer to the network. This would involve augmenting the existing traffic or demand matrix with estimates of the traffic volumes introduced by the new customer.

2.3 PRACTICAL CHALLENGES

Several comments about the three canonical representations are in order. First, these representations rely on a decoupling into an *invariant* external or offered load, described by the traffic and demand matrices, and the spatial flow of the traffic resulting from this load, described by the path matrix. This decomposition is an idealized assumption. In fact, the Internet has many closed control loops that *adapt* the load to the network behavior itself. For example, TCP congestion control throttles the transmission rate in response to packet loss on a relatively small time scale; content distribution networks (CDNs) may redirect requests for web pages to alternate servers in response to network conditions; and end users themselves adapt their behavior, perhaps abandoning a web transfer if the response time is unacceptably long. Therefore, the decoupling inherent in the representations described above is a simplification, and the precision of predictions based on such a load characterization may have some inherent inaccuracy.

Second, the path, traffic, and demand matrices can be very large objects, especially in a large IP backbone. For example, the set of all ingress-egress pairs is on the order of n^2 in a network with n nodes; the path matrix can potentially be very much larger, depending on the frequency of route changes and the existence of multiple shortest paths between ingress-egress nodes. The demand matrix could be even larger, depending on the number of different egress sets that arise in practice. Thus, the representations as described here are mainly conceptually appealing; in most practical settings, they would not actually be fully instantiated, but only populated partially, depending on the specific operational applications they drive (e.g., DDoS attack detection, service-level agreement (SLA) verification, routing). Defining compact and useful partial representations for specific applications is an important practical challenge, which has to rely on a diverse set of expertise and technology, such as databases, statistics, and algorithms.

Third, we have simplified the discussion of the path, traffic, and demand matrices in that we have implicitly assumed that the only attribute of interest is the traffic volume as a function of time. In the simplest case, the traffic volume represents the average load over some time scale (e.g., on a per-hour basis); however, other bandwidth measures, such as peak rate or effective bandwidth [20], are also possible. In practice, the network operators may need to focus their analysis on a particular subset of traffic, such as the path matrix for web traffic or a particular customer. Alternatively, the operators may want to have a separate element in the traffic for each application (say, based on the TCP port numbers). This adds another dimension to these representations, reinforcing our point that efficient storage and querying of such large, high-dimensional objects is a challenging and important research problem.

3   PASSIVE MEASUREMENT OF IP TRAFFIC

This section reviews the state-of-the-art in techniques for collecting traffic measurements in large IP backbone networks. We focus on passive measurement techniques that accumulate information about the traffic as it travels through the network. First, we discuss how operators can collect aggregate traffic statistics using SNMP and highlight some of the advanced features available for local-area networks (LANs) using RMON. Next, we describe how operators can collect detailed IP packet traces on individual links in the network. Then, we describe techniques for monitoring traffic at the flow level to provide relatively detailed traffic statistics with fewer measurement records. Throughout the section, we consider the limitations of each measurement technique in providing a detailed, network-wide view of the prevailing traffic. Also, although the main focus of this chapter is on measurements to populate domain-wide traffic representations, each subsection briefly discusses other, more direct, uses of the different types of passive measurement data.

3.1 SIMPLE NETWORK MANAGEMENT PROTOCOL/REMOTE MONITORING

The simple network management protocol (SNMP) is an Internet engineering task force (IETF) standard for the representation of management information, and the communication of such information between management stations and management agents for the purpose of monitoring and affecting network elements [34, 35]. Both the structure of the management information base (MIB) and the communication protocol itself are kept as simple as possible: MIB information uses a hierarchical naming structure, where a leaf of the naming tree can be one of only two types: a simple scalar variable or a table. The communication protocol supports only three primitives: get (i.e., reading of an agent variable by the management station), set (i.e., changing such a variable by the management station), and get-next (i.e., reading the next item in lexicographical order, which facilitates reading a table). There is also support in the protocol for automatic notifications from the agent to the management system. The first version of SNMP has since been extended in various ways; some of its original simplicity has inevitably been lost in the process [33].

The main standardized MIB containing traffic-related data is called MIB-II. It is organized in a set of groups (subtrees) that monitor the execution of various protocols on the network element, such as IP, TCP/UDP, BGP, OSPF, etc. Most of this information is either status information (e.g., the operational status of the interfaces on a router), or highly aggregated (e.g., per-interface counters of bytes and packets inbound and outbound). There is no support for fine-grained measurements required to populate the representations presented in the previous section.

The advantage of MIB-II is that it is almost universally supported in routers and other network elements, even on high-speed interfaces. A drawback lies in the absence of a certification authority to ensure correct implementations. Unfortunately, many vendors seem to do an insufficient job in testing MIB-II implementations, which results in measurement errors and resultant artifacts.

RMON is another standardized SNMP MIB, whose goal is to facilitate remote monitoring of LANs [29]. Its main advantages are (1) to enable a single agent to monitor an entire shared LAN; (2) to endow the agent with local intelligence and memory to enable it to compute higher-level statistics and to buffer these statistics in the case of outages; (3) to define alarm conditions and the actions that should be taken in response, such as generating notifications to the network management system; and (4) to define packet filtering conditions and the actions that should be taken in response, such as capturing and buffering the content of these packets.

RMON offers great flexibility in combining the above primitives into sophisticated agent monitoring functions. However, this flexibility makes a full RMON agent implementation costly. Thus, RMON has only been implemented (at least partially) for LAN router interfaces, which are relatively low speed. Implementations for high-speed backbone interfaces have proved to be infeasible or prohibitively expensive. Instead, router vendors have opted to develop more limited monitoring capabilities for high-speed interfaces. The next two subsections describe the two main classes of such methods, *packet monitoring* and *flow measurement*. v

Direct uses. MIB-II information is useful to verify the overall operational health of a network, by monitoring variables related to traffic (e.g., link utilization, fraction of packets dropped due to checksum errors), router health (e.g., CPU and memory load of the router's central processor), and network state and protocol performance (e.g., resets on BGP sessions). Such variables are typically monitored continuously, but on a relatively slow time scale (e.g., minutes to hours), by a central network management system. This management system issues alarms to human operators if monitored variables are outside their predefined ranges. For many failure scenarios, however, MIB-II information is too highly aggregated to be sufficient for diagnosis. For example, if a link is overloaded due to a denial-of-service attack, the operator requires more fine-grained traffic measurements (such as packet or flow measurements) to determine the origin of the attack.

RMON gives access to much more fine-grained measurements, and provides more flexibility in how these measurements are collected. For example, it enables a network element to collect a traffic matrix at the MAC-address (the layer-2 medium access control address) level directly and to report those host pairs that exchange the most traffic to the network management system. Such a functionality would be very useful in the example of a denial-of-service attack. RMON also allows the agent to accumulate a time series of samples of a variable, and to perform limited local processing on this time series (e.g., alarm thresholding). RMON embodies support for monitoring of packets that match a filter criterion; fields of interest of these packets can be reported back to the network management system.

In fact, if RMON were universally deployed on high-speed interfaces, it would essentially obviate the need for additional packet and flow monitoring support. However, its complexity appears to prohibit implementation other than in LAN environments, which relegate RMON to the edge of the network. It is not widely used by the operators of large backbone networks.

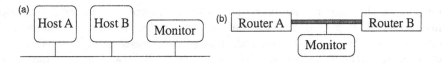

FIGURE 4   Tapping a link to collect packet traces. (a) Shared media and (b) point-to-point link.

3.2 PACKET MONITORING

A packet monitor passively collects a copy of the packets traversing a link and records information at the IP, TCP/UDP, or application layers. Collecting packet traces requires an effective way to tap a link in the operational network. This is relatively easy in a local area network (LAN) consisting of a shared medium, such as an Ethernet, an FDDI ring, or a wireless network. As shown in figure 4(a), the packet monitor may be a regular computer connecting to the LAN with a network interface card configured to run in *promiscuous* mode to make a local copy of every packet. Tapping a link is more difficult in a large backbone consisting of routers connected via high-speed point-to-point links. Traces may be collected by splitting each unidirectional link into two parts in order to direct a copy of each packet to the monitor, as shown in figure 4(b). Alternatively, the router could include support for collecting packet traces. As traffic reaches the router, the interface card can make a local copy of the packets.

Copying and analyzing the entire contents of every packet is extremely expensive, especially on the high-bandwidth links common in large backbone networks. Limiting the processing load and the volume of measurement data is crucial. Packet monitors employ three main techniques to reduce the overhead of collecting the data. First, the monitor can capture and record a limited number of bytes for each packet. For example, the monitor could record the IP header (typically 20 bytes) or the IP and TCP headers (typically 40 bytes) instead of the entire contents of the IP packet. Second, the monitor may be configured to focus on a specific subset of the traffic, based on the fields in the IP and TCP or UDP headers. For example, the monitor could capture packets based on the source and destination IP addresses or port numbers. Third, the monitor may perform sampling to limit the fraction of packets that require further processing. For example, the monitor could be configured to record information for one out of every hundred packets. Routers that support packet-level measurement typically employ sampling to reduce the overhead on the interface card and avoid degrading packet-forwarding performance.

Packet monitoring is an effective way to acquire fine-grained information about the traffic traversing a link. PC-based monitoring platforms are quite common in LAN environments. Many of these packet monitors run the popular, public-domain tcpdump software [22] using the Berkeley packet filter [26]. Extensions to this software support the collection of application-level information inside the IP packets to analyze web or multimedia traffic or detect possible intruders [12, 28, 38]. However, collecting packet traces on high-speed links is challenging without dedicated hardware support. Several monitors have been developed by research groups to collect traces on selected links [3, 18, 23]. In addition, several companies offer packet monitoring products that can collect traces on a variety of different types of links. Network operators can use these products to collect traces on key links that connect to important customers or services. Recognizing the importance of fine-grained traffic measurements to service providers, major router vendors have begun to support packet sampling directly in their interface cards and the IETF has started to define a small set of standard features [10].

Direct uses. A packet monitor can provide network operators with detailed information about the traffic traversing a link. First, the traces can be aggregated to generate reports of the volume of traffic in terms of key fields in the packet headers. For example, computing traffic volumes by source and destination IP addresses allows operators to identify the end hosts responsible for the bulk of the load on the network. Considering the traffic mix by TCP or UDP port number is useful for identifying the most popular applications (e.g., web, peer-to-peer transfers, etc.). Second, packet traces can provide information about the performance experienced by users. For example, the traces can be analyzed to compute the throughput for individual TCP connections or source-destination pairs. With accurate timestamps, the operators can combine traces from multiple locations in the network to compute latency statistics. Flags in the TCP header may be useful for identifying when a user aborts an ongoing transfer or for detecting certain kinds of denial-of-service attacks. Third, the traces can provide insight into the fine-grained characteristics of the traffic, such as packet sizes, the burstiness of the stream of packets traversing the link, and typical transfer sizes.

3.3 FLOW MEASUREMENT

Flow-level measurement involves collecting aggregate traffic statistics for related packets that appear close together in time. The abstraction of a flow mimics the notion of a call or connection in circuit-switched networks. Since IP networks do not retain information about individual data transfers, the grouping of packets into a flow depends on a set of rules applied by the measurement device. The rules identify the attributes that must match across all the packets in a flow, as well as restrictions on the spacing between packets in a flow. For example, a flow could be defined as all packets that have the same source and destination IP addresses, where successive packets have an interarrival time of less than 30 seconds. Alternatively, a flow could be defined as all packets that match in their IP addresses, port numbers, and protocol, and have an interarrival time less than 60 seconds. The rules for defining a flow depend on the details of the

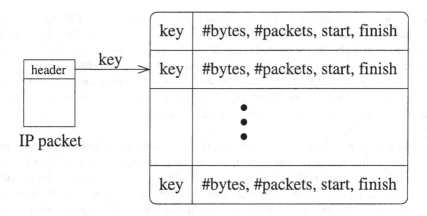

FIGURE 5   Accessing the flow cache with a key computed from the packet header.

measurement device and may be configurable. The measurement device reports aggregate information for each flow, such as the key IP and TCP/UDP header fields that match across the packets, the total number of bytes and packets in the flow, and the timestamps of the first and last packets.

Similar to a packet monitor, a flow-measurement device must tap a link to receive a copy of each packet. To collect the aggregate traffic statistics, the device maintains a cache of active flows. When a packet arrives, the device computes an index into the cache by combining the various header fields that define a flow, as shown in figure 5. The device accesses the cache to create a new entry or update an existing entry. Each entry in the cache includes the aggregate traffic statistics, such as the total number of bytes and packets, as well as the timestamps. The device evicts an entry from the cache and exports or records the information. A flow is evicted from the cache if no additional packets have arrived for a period of time (e.g., 30 or 60 seconds) or if the cache is full. The device may also evict a flow that has been active for a long period of time (e.g., 30 minutes) to ensure that information about the long-lived transfer is made available. The arrival of additional packets with the same index would trigger the creation of a new entry in the cache and, ultimately, another measurement record. In practice, the device may combine multiple such records into a single unit for transmission to a separate machine for archiving and analysis.

Flow-level measurements are available from some router vendors, most notably Cisco with its Netflow feature [5]. For manageable overhead on high-speed links, recent releases of Netflow include support for sampling and aggregation [6, 7]. Several commercial packet monitors can produce flow-level measurements, allowing network operators to have a mixture of router support and separate measurement devices. The IETF has made efforts to define standards for flow-level measurement. The real-time traffic flow meter (RTFM) group [2, 19] provided a formal definition of flows and described techniques for collecting the measurement data. More recently, the IP flow information export (IPFIX) working group [21] is defining a format and protocol for delivering flow-level data from the measurement device to other systems that archive and analyze the information. The vendor and standardization activity reflects a growing interest in the use of flow-level measurement as a practical alternative to collecting packet traces.

Direct uses. Flow-level measurements are useful for many of the same purposes as packet traces. A network operator may use flow-level statistics to compute traffic volumes by IP addresses, port numbers, and protocols, or basic traffic characteristics such as average packet size. An ISP may use traffic statistics at the level of individual customers to drive accounting and billing applications. However, flow-level measurements do not provide the fine-grained timing information available in packet traces. The flow measurement record provides timestamps for the first and last packet in the flow, without detailed information about the spacing between successive packets. As such, flow-level measurements are not useful for studying the burstiness of traffic on a small time scale. Flow-level measurements can provide some information about the throughput experienced by users, based on the number of bytes and the time duration of a flow. However, identifying potential causes of low throughput, such as lost and retransmitted packets, is difficult without more detailed information.

4   POPULATING THE NETWORK-WIDE TRAFFIC MODELS

This section describes how the different types of measurement data discussed in section 3 can be used to populate the network-wide traffic models described in section 2. We present these methods in a progression from coarse-grained traffic statistics to fine-grained packet traces to illustrate how a finer granularity of measurement data allows us to relax our assumptions about the traffic and network models. In particular, we discuss the following three scenarios:

- **Assumptions on the network and traffic model:** First, we assume that only aggregate traffic statistics, such as traditional SNMP data, are available for each link. In this scenario, a class of statistical inference methods referred to as network tomography can generate an *estimate* of the traffic matrix. The inference techniques rely on assumptions about the state of the network elements and the statistical properties of the traffic.
- **Assumptions on the network model:** Second, when more fine-grained traffic measurements, such as flow or packet traces, are available from the links at the edge of the network, the traffic and path matrix can be *derived*. In contrast to network tomography, this approach does not rely on any assumptions about the characteristics of the traffic. However, the technique does need to model

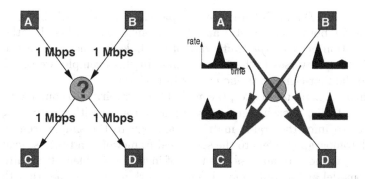

FIGURE 6 Left: a single observation of aggregate link rates provides no information about the traffic matrix. Right: multiple observations of the link rates through repeated measurements over time, on the other hand, enables the inference of the traffic matrix. Intuitively, when the aggregates rates on two links are strongly correlated, we expect these two links to carry the same traffic flows.

certain aspects of the network in order to "map" the edge measurements onto the topology.

- **No assumptions about network or traffic:** Third, we describe methods that directly *observe* the path matrix without relying on any assumptions about the traffic or the network. These methods yield the most faithful observation of the spatial distribution of traffic even in scenarios where the network state is unknown or changing. However, these techniques require additional support in routers that is not available in commercial products as of this writing.

4.1 NETWORK AND TRAFFIC MODEL: TOMOGRAPHY

The class of methods we describe in this section apply in a setting where only aggregate traffic measurements are available. Specifically, these methods assume that the data rate on every link in the domain is known, averaged over some time interval. The goal is to estimate the traffic matrix, that is, the data rate for every ingress-egress pair, without any explicit packet or flow measurements.

The basic idea of network tomography is quite intuitive. Consider the simple network in figure 6, consisting of four ingress-egress pairs AC, AD, BC, and BD, and four links, on which the aggregate rate is measured. It is quite clear that with a single measurement, nothing can be said about the traffic matrix. For example, if the data rate on each link were 1 Mbps, this does not tell us the relative contributions of AC and AD on the upper left link. However, if the evolution over time of link data rates is known, then the traffic matrix can be estimated. For example, given the link rates in figure 6(b), we would intuitively conclude that AD and BC are large compared to AC and BD, because the rate functions for both links on AD and BC resemble each other. In other words, the correlation between the traffic rate of two links is a measure for the amount of traffic that traverses both of these links. In this example, we could directly estimate the four-element traffic matrix in this way.

In more general topologies, the goal of tomography is to combine knowledge of how the traffic of each ingress-egress pair is routed with measured aggregate link rates to infer the traffic matrix in analogy to the simple scenario above. Network tomography refers to the statistical framework and the associated computational methods to address this type of inversion problem. It requires both a network model and a traffic model. The network model assumes that the routes from every ingress to every egress pair are stable and known over a time period over which inference is performed. The traffic model describes traffic flows as some parametrized stochastic process. Such a traffic model is central to tomography methods, as they inherently rely on multiple measurements over time to infer the missing spatial information about traffic flow.

The pioneering work in this area is due to Vardi [39], and inspired by conceptually similar prior work in transportation networks. We give an overview of his model and of his approach to estimating the traffic matrix. The subsequent discussion requires that we introduce some notation. Let the vector $Y = (Y_1, \ldots, Y_r)$ denote an *observation* of packet counts on all links, where r is the number of links in the network, and Y_i is the number of packets observed on link i. The vector $X = (X_1, \ldots, X_c)$ counts the number of packets per ingress-egress pair, where c, the number of pairs, is typically much larger than r. X can be viewed as a snapshot of the traffic matrix. The relationship between X and Y can be written in matrix form as

$$Y = AX \qquad (1)$$

where A is the routing matrix, embodying the fixed and known routing state as follows: the element $A_{i,j} = 1$ if link i is on the path of ingress-egress pair j, and 0 otherwise.

In general, the number of ingress-egress pairs is much larger than the number of links in the network (i.e., $c \gg r$). As such, a single observation of Y contains very little information about X, in the sense that the subspace $\{X : Y = AX, X \geq 0\}$ of possible X for a given observation Y is of high dimension. Vardi, therefore, introduces the following parametrized stochastic model. Assume that time is slotted, and that for each time slot k, we obtain an independent sample of the traffic matrix $X^{(k)}$, and hence of the observed link rates $Y^{(k)}$. For each k, the components of $X^{(k)}$ are independent and have Poisson distribution with mean $\lambda = (\lambda_1, \ldots, \lambda_c)$,

$$X_j^{(k)} = \text{Poisson}(\lambda_j). \qquad (2)$$

Let us consider the maximum likelihood estimator $\hat{\lambda}$ of λ. It is relatively straightforward to write down the likelihood function for λ,

$$L(\lambda) = P_\lambda(Y) = \sum_{X:AX=Y,X\geq 0} P_\lambda(X). \tag{3}$$

Unfortunately, the likelihood function is very expensive to compute, because it involves finding all natural solutions to the equation $AX = Y$. Vardi proposes an estimator based on the method of moments, which is very efficient to compute. Specifically, for large sample size K, the empirical mean $\bar{Y}$ of the observation Y, given by

$$\bar{Y} = \frac{1}{K} \sum_{k=1}^{K} Y^{(k)}, \tag{4}$$

is approximately Gaussian. Therefore, its distribution depends only on its first and second moments. By equating the empirical moments of $\bar{Y}$ with the moments predicted by the model, we obtain the equation

$$\begin{pmatrix} \bar{Y} \\ S \end{pmatrix} = \begin{pmatrix} A \\ B \end{pmatrix} \lambda s. \tag{5}$$

Here, S is the empirical covariance matrix, written as a vector, and the matrix B is a function of A. The linear system (5) is, in general, inconsistent because of the noise in the empirical moments, and cannot be solved directly. Vardi proposes heuristics to eliminate equations from (5), and to iteratively compute an approximate estimator $\hat{\lambda}$.

Several papers build on and extend Vardi's work. For example, Tebaldi and West [37] study the tomography problem in a Bayesian setting, which allow for the inclusion of prior information into the estimate. Cao et al. [4] consider a similar problem setting, but instead of the (discrete) Poisson traffic model, they assume that the data rate for every ingress-egress pair is Gaussian. Their model is more flexible than Vardi's because it parametrizes the scaling relationship between the mean and variance of the Gaussian marginals.

Network tomography methods have some serious practical limitations. First, they tend to make strong simplifying assumptions about the traffic model in order to devise tractable solutions. The assumptions typically include the stationarity and ergodicity of the traffic process in order to estimate the spatial traffic matrix from multiple temporal samples. Furthermore, it is usually assumed that consecutive samples are independent. However, network traffic tends to possess much more complicated structure [25], and can be viewed as stationary only over relatively short periods of time; also, consecutive samples will not typically be completely independent. Second, even when the traffic model is accurate, it takes a large number of samples to arrive at sufficient precision. This is a direct consequence of the fact that in most cases, the number of unknowns (the elements of the traffic matrix) is much larger than the number of observables (the aggregate rate on links).[3] Third, given the size of the inversion problem (5), estimating a large traffic matrix is computationally challenging. An interesting research question, therefore, concerns the inference of a subset of the traffic matrix, depending on the application it is driving.

These limitations suggest that tomography methods are not applicable to large-scale traffic management problems. However, they may very well turn out to be useful in limited applications. For example, suppose an operator wishes to obtain a "local" traffic matrix at a single router. Tomography may work well in such a limited setting, and could turn out to be an interesting alternative to collecting detailed packet or flow traces from that router. Another promising approach of recent interest reduces the number of unknowns by making additional assumptions about the form of the traffic matrix. For example, Medina et al. [27] and Roughan et al. [30] adopt *gravity models*, where each ingress and egress node is assumed to possess a "mass." The amount of traffic between an ingress and an egress node is then assumed to be proportional to the product of the respective masses, similar to the force of gravity between two physical points of mass. This assumption dramatically reduces the number of unknowns (from $O(c^2)$ to $O(c)$). However, further study is required to determine to what extent gravity models fit real traffic matrices. Finally, a more limited use of tomography consists in complementing other types of traffic measurements. For example, if, in a network domain, only a subset of edge routers were capable of obtaining packet or flow measurements, then tomography methods might be useful for inferring the missing components of the traffic matrix. While such a complementary use of tomography is promising, it requires further research.

4.2   NETWORK MODEL: TRAFFIC MAPPING

Developing simple and accurate statistical models of Internet traffic is difficult in practice. Packet and flow-level measurements offer a way to construct the path, traffic, and demand matrices without simplifying assumptions about the characteristics of the traffic. However, collecting fine-grained traces for every link in a large network would be expensive given the limitations of today's technology. Even if the interface cards have direct support for traffic measurement, this functionality may not be uniformly available and, in some cases, may degrade the throughput of the router. In addition, many routers do not provide effective, tunable ways to limit the volume of measurement data. Given these constraints, it is appealing to have a way to construct the path, traffic, and demand matrices based on a limited collection of fine-grained measurements. For example, monitoring every ingress point into the domain provides a way to collect fine-grained statistics for all of the traffic without enabling measurement functionality in the core of the network. However, computing the spatial distribution of traffic

---

[3]Vardi presents some numerical examples using a 5-node topology that illustrate this problem.

from these measurements depends on having an accurate model of routing in the underlying network.

Populating the network-wide models involves collecting fine-grained measurements at each ingress point and aggregating this data to the level of the path, egress point or the set of possible egress points. The forwarding of traffic through the network depends on the IP prefix associated with each packet's destination IP address. Each router combines information from the intradomain and interdomain routing protocols to construct a forwarding table that is used to select the next-hop interface for each incoming packet. For example, a forwarding table has entries such as:

12.128.0.0/9	10.126.212.1	POS2/0
172.12.4.0/24	10.1.2.118	Serial1/0/0:1
192.0.2.0/24	10.47.35.56	POS3/0

The third entry directs packets destined for 192.0.2.0/24 to interface POS3/0 (a packet-over-SONET link in slot 3 of the router) to reach the next hop with IP address 10.47.35.56. When a packet arrives, the router performs a longest prefix match on the destination address to find the appropriate forwarding-table entry for the packet. For example, the router could associate the IP address 192.0.2.38 with the prefix 192.0.2.0/24.

Most routers have a command-line interface that allows network operators to view the current state of the forwarding table; operators may run scripts that dump these tables periodically. These dumps provide the list of prefixes and their associated interfaces. The routers that connect to neighboring domains have forwarding tables that indicate the egress point(s) associated with each destination prefix. For example, suppose the Serial1/0/0:1 interface (in the forwarding table listed above) connects to a customer's network; then, the destination prefix 172.12.4.0/24 would have this interface in its set of egress points. Other routers might have forwarding table entries that direct traffic to an interface to another router inside the AS; these interfaces would not be included in the egress set. Constructing the set of egress interfaces for a destination prefix involves extracting the entries from each forwarding table and identifying the next-hop interfaces that are associated with neighboring domains. Based on this information, the fine-grained traffic measurements collected at the ingress points can be associated with the longest matching destination prefix and, in turn, the set of egress points [14].

On the surface, the demand matrix may seem like an unwieldy representation that has a large number of elements. A typical forwarding table in a large IP network contains more than 100,000 prefixes, and this number is increasing as the Internet continues to grow. In addition, the number of egress sets could be exponential in the number of distinct egress points. In practice, however, many destination prefixes have the same egress set. Many organizations, such as companies or universities, announce multiple blocks of IP address to their Internet service providers. Another AS in the Internet would tend to route all traffic for

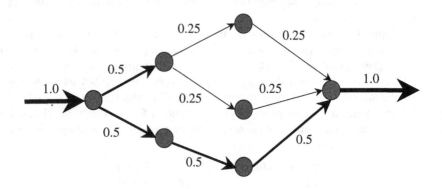

FIGURE 7   Traffic splitting across multiple shortest paths.

these destination prefixes in the same manner, since the BGP advertisements would have all the same attributes. In addition, most combinations of egress points do not arise in practice. Typically, an egress set would consist of all links to a particular neighboring domain, such as a customer or peer. For example, a network may connect to another AS in six different locations. These six links would tend to appear together in egress sets for various destination prefixes. This tends to result in a relatively small number of different egress sets, ensuring a compact representation for the demand matrix.

The demand matrix can be used to compute the path and traffic matrices. Each element in the demand matrix corresponds to a single ingress point and set of egress points. The traffic from that ingress point travels to a particular egress point over one or more paths through the domain. The selection of a particular egress point depends on the intradomain routing configuration (e.g., OSPF or IS-IS weights), as discussed earlier in section 2.1. The chosen egress point and the path(s) through the domain can be determined in two main ways. First, by dumping the forwarding table for each router in the domain, it is possible to identify the interfaces that would carry the packets as they travel from one router to another. Second, these paths can be computed based on a network-wide view of the topology and the intradomain routing configuration (e.g., by computing the shortest path(s) based on the OSPF/IS-IS weights) [13]. In practice, operational networks may have multiple shortest paths from the ingress point to the chosen egress point. Most commercial routers split traffic over these multiple paths, dividing the traffic evenly when multiple next-hop interfaces lie along a shortest path, as shown in figure 7.

4.3   NO MODEL: DIRECT OBSERVATION

The methods presented so far are all *indirect* to a certain degree, in that they rely on some assumptions on the traffic and on the network behavior to populate the various domain-wide measurement representations. These assumptions are embodied in the traffic and network models. Indirect measurement methods thus suffer from the inherent uncertainty associated with the physical and logical state of large, heterogeneous networks. This uncertainty has several sources:

- The exact behavior of a network element, such as a router, is not exactly known to the service provider and depends on vendor-specific design choices. For example, the algorithm for splitting traffic among several shortest paths in OSPF is not standardized.
- The network has deliberate sources of randomness to prevent accidental synchronization, e.g., through randomized timers in routing protocols [16].
- IP packets may be lost as they travel through the network. Predicting whether a packet measured in one location was successful in traversing the entire path through the domain is difficult in practice.
- Some of the behavior of the network depends on events outside of the control of the domain. For example, how traffic is routed within an autonomous system (AS) depends in part on the dynamics of BGP route advertisements from neighboring domains [24].
- The interaction between adaptive schemes operating at different time scales and levels of locality (e.g., QoS routing, end-to-end congestion control) may simply be too complex to characterize and predict [17].
- Equipment failures can disrupt the normal operation of the network. Traffic measurement is one of the tools for detecting and diagnosing such problems; however, this benefit is mitigated if traffic measurement depends on the correct operation of the network.

Rather than relying on a network model and an estimation of its state and its expected behavior, *direct* methods observe the traffic as it travels through the network. In this subsection, we describe two techniques for direct observation of the path matrix—trajectory sampling and IP traceback. Trajectory sampling is a method for network-wide packet sampling; conceptually, it directly provides an estimator of the path matrix. IP traceback is also a sampling-based method, but it is tailored to a particular traffic management application: the detection of distributed denial of service (DDoS) attacks. This can be viewed as estimating a particular slice of interest of the full path matrix.

4.3.1   Trajectory Sampling.

Trajectory sampling [9] involves sampling packets that traverse each link (or a subset of these links) within a measurement domain. The subset of sampled packets over a certain period of time can then be used as a representative of the overall traffic.

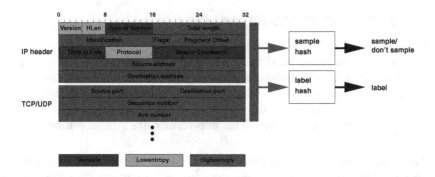

FIGURE 8 Trajectory sampling relies on two hash functions computed over the invariant fields of the packet's header and payload. The sampling hash function serves to decide whether a packet should be sampled or not; the label hash function stamps a unique identity on the packet for trajectory reconstruction.

If packets were simply randomly sampled at each link, then we would be unable to derive the precise path that a sampled packet has followed through the domain from the ingress to the egress point. The key idea in trajectory sampling is, therefore, to base the sampling decision on a deterministic hash function over the packet's *content*. If the *same* hash function is used throughout the domain to sample packets, then we are ensured that a packet is either sampled on *every* link it traverses, or on no link at all. In other words, we are effectively able to collect *trajectory samples* of a subset of packets. The choice of an appropriate hash function will obviously be crucial to ensure that this subset is not statistically biased in any way. For this, the sampling process, although a deterministic function of the packet content, has to resemble a random sampling process.

A second key ingredient of trajectory sampling is that of *packet labeling*. Note that to obtain trajectory samples, we are not interested in the packet content per se; we simply need to know that *some packet* has traversed a set of links. But to know this, it is sufficient to obtain a unique packet identifier, or label, for each sampled packet within the domain and within a measurement period. Because the label is unique, we will know that a packet has traversed the set of links which have reported that particular label. Trajectory sampling relies on a second hash function to compute packet labels that are, with high probability, unique within a measurement period. While the size of the packet labels obviously depends on the specific situation, it has been shown that labels can, in practice, be quite small (e.g., 20 bits). As the measurement traffic that has to be collected from nodes in the domain only consists of such labels (plus some auxiliary information), the overhead to collect trajectory samples is small.

Trajectory sampling has several important advantages. It is a direct method for traffic measurement, and, as such, does not require any network status in-

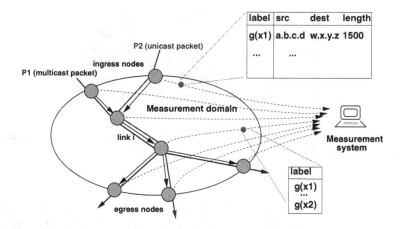

FIGURE 9   Schematic representation of trajectory sampling. A measurement system collects packet *labels* from all the links within the domain. Labels are only collected from a pseudo-random subset of all the packets traversing the domain. Both the decision whether to sample a packet or not, and the packet label, are functions of the packet's invariant content.

formation. A set of trajectory samples is basically a sample of the path matrix. Trajectory sampling does not require router state (e.g., per-flow cache entries) other than a small label buffer. The amount of measurement traffic necessary is modest and can be precisely controlled. Multicast packets require no special treatment—the trajectory associated with a multicast packet is simply a tree instead of a path. Finally, trajectory sampling can be implemented using state-of-the art digital signal processors (DSPs) even for the highest interface speeds available today.

4.3.2 IP Traceback.   The second domain-wide measurement method that does not rely on a network model or a traffic model is *IP traceback* [31]. The main goal of IP traceback is detecting and diagnosing distributed denial of service (DDoS) attacks. In such an attack, a potentially large number of hosts (which may have been subverted) overwhelm a victim host, such as a web server, with so much traffic that it becomes inoperable. Defending against such an attack is difficult, because the source addresses of the attack packets are often *spoofed*, that is, they do not correspond to the real originator of the packet. Therefore, such packets contain no information about the originator. The goal of IP traceback is to infer the originator(s) of such an attack, or to at least partially infer the paths from the originator to the victim. This allows operators to take countermeasures, such as blocking traffic on certain links, and isolating "hijacked" hosts used in the attack.

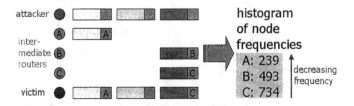

FIGURE 10   Node sampling. The attack path is reconstructed simply through the observed frequency of node identities.

IP traceback is an end-to-end and interdomain solution to this problem. It is end-to-end because the victim does not require any explicit help from its network operator to infer the attack sink tree. It is interdomain because no specific coordination across domain boundaries is required to infer a sink tree spanning multiple domains. These properties enable a gradual and widespread deployment of IP traceback.

We describe IP traceback by first introducing a very crude scheme to solve this inference problem, and then successively refining it. A straightforward solution to the traceback problem would consist in having every router add its identity to a list carried with the packet. This would allow the recipient of the packet to determine the exact path followed by the packet. Of course, such a scheme would incur prohibitive processing and communication overhead.

To limit this overhead, it is desirable to allocate only a small, fixed field in each packet to IP traceback. Clearly, this prohibits fully reporting the path traversed by each packet. A possible solution consists in having packets sample the set of routers encountered along their path, and using the field to report the last sampled router to the recipient (cf. fig. 10). In practice, this can be implemented by routers flipping a coin for each packet, and writing their IP address into the traceback field with probability p. As the defining property of a denial of service attack is that the victim receives a large number of attack packets from the same host or set of hosts, the victim can construct a histogram of router addresses observed in received packets. In the idealized case of a single attacker and attack path, the observed frequencies of router addresses directly yield the set of routers on the path, the most frequently observed address corresponding to the router closest to the attacker, and vice versa.

The above approach has two problems. First, it is possible for the attacker to perturb the inferred attack path by preloading the traceback field of all the packets it sends with some forged IP address (or possibly several addresses). This forged address will show up in the frequency histogram; the victim has no way to tell forged and real samples apart. The only way to avoid this effect would be by using a very high sampling rate of $p > 1/2$. However, this would preclude inferring reasonably long paths, as the probability for a sample to survive decreases very

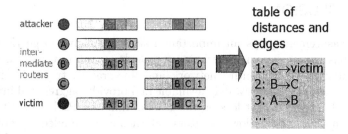

FIGURE 11 Edge sampling. The attack path or tree is explicitly reconstructed from sampled edges.

quickly with router distance from the victim. Second, the node-based scheme cannot distinguish multiple attack paths, as the path is simply inferred from sample frequencies.

To remedy these two problems, the scheme can be slightly extended to sample edges instead of nodes. This can be achieved through three fields, **edge_start**, **edge_end**, and **dist** (cf. fig. 11). Each router again flips a coin with probability p. In case of a hit, it writes its address into the field **edge_start**, and sets **dist** to zero. The next router, because it receives a packet with **dist**=0, writes its identity into **edge_end**, and increments **dist**. Every downstream router will simply increment **dist** (provided it does not start an edge sample itself). Therefore, the receiver will effectively receive a sampled edge (**edge_start**, **edge_end**), along with the distance **dist** of that edge from itself.

It is straightforward to see that the attack sink tree can be fully reconstructed from edge samples. However, the edge-based scheme just described would require 64 bits for **edge_start** and **edge_end**, plus the bits required for **dist**. The main challenge addressed in Savage et al. [31] is how to compress this information so that it fits into the 16-bit identification field of the IP header. The authors propose several clever tricks to achieve this compression.

Related proposals for the diagnosis of DDoS attacks rely on out-of-band methods (specifically, ICMP packets) to carry samples [1], and *hash-based IP traceback* [32], a method that keeps compressed path information within routers. The latter method is similar to trajectory sampling, in that it relies on a hash computed over invariant packet fields to associate unique labels with packets. In contrast to trajectory sampling, however, the method keeps track of all the packets traversing a router.

5 CONCLUSION

Traffic measurement plays an important role in managing large IP networks. Existing passive measurement techniques provide operators with coarse-grained traffic statistics for the entire network and fine-grained traces for individual links and routers. SNMP is widely available and provides aggregated information about link load and packet loss on the scale of minutes. Packet monitoring and flow-level measurements allow operators to have a more detailed view of the traffic; however, these techniques are not uniformly available in existing IP routers. Operators need a network-wide view of the spatial distribution of traffic to detect and diagnose performance problems, and to evaluate potential control actions. The path, traffic, and demand matrices provide a sound basis for a variety of common tasks, such as identifying the customers affected by an overloaded link and tuning the routing configuration to alleviate the congestion.

Populating these traffic models requires collecting measurement data from multiple locations in the network. The path and traffic matrices can be estimated from traditional SNMP measurements, under certain assumptions about the statistical properties of the traffic and knowledge of routing inside the backbone. Collecting fine-grained packet or flow-level measurements at the edge of the network makes it possible to compute the path, traffic, and demand matrices without any simplifying assumptions about the traffic. With additional support for packet sampling in the routers, the path matrix can be observed directly without depending on additional information from the routers.

The existing work on these network-wide representations suggests several avenues for future research:

- **Trade-off between efficiency and accuracy:** Populating the traffic models involves collecting a large amount of measurement data from multiple locations in the network. The accuracy of these models depends on the assumptions underlying the techniques and the granularity of the measurement data. Although accuracy may be crucial for some applications (such as billing), rough estimates of the path, traffic, and demand matrices may be sufficient for many important tasks (such as selecting routing parameters to balance load in the network). Future work should identify how well certain applications can tolerate inaccuracy in the traffic estimates and explore how to exploit these observations to reduce the measurement overhead, for example, through sampling techniques.
- **Uniform support for traffic measurement:** Support for traffic measurement has traditionally been an afterthought in the design of IP routers. Although SNMP is a standard protocol, most routers provide relatively limited, coarse-grained load statistics; the fine-grained measurements defined in RMON are not available on high-speed backbone links. Instead, vendors selectively provide proprietary support for packet sampling and flow-level measurement. Ultimately, IP equipment needs to evolve to include more advanced

measurement functionality in a uniform manner. The recent IETF activities surrounding packet sampling [10] and flow export [21] are positive steps in this direction. Future work should consider effective hardware support for collecting fine-grained measurements on high-speed links and ways to estimate important statistics about the traffic based on sampled data.

- **Network-wide models of the traffic:** The path, traffic, and demand matrices are helpful abstractions that can drive a variety of important tasks in operating a large IP network. In practice, individual applications may warrant tailored ways of populating (parts of) these views. For example, although the complete path matrix may be a very large data structure, traffic statistics for a small subset of these paths may be sufficient to diagnose a DoS attack. The existing models focus primarily on operational tasks within a single network, without considering how the traffic flows through the rest of the Internet. In practice, the Internet is an extremely complex "network of networks," and most traffic must traverse several domains en route to its destination. While there is no hope of modeling this large-scale complex system exhaustively, new models should incorporate enough detail to predict the impact of domain-local control actions. As these control actions become more wideranging and sophisticated (e.g., traffic differentiation, explicit routing through MPLS, failover protection), more fine-grained traffic models will become necessary. Specifically, future work should investigate models that capture the influence of operator actions on neighboring domains.

Progress in these areas of research would be extremely valuable in helping operators run their networks more efficiently.

ACKNOWLEDGMENTS

We would like to thank Shubho Sen and Yin Zhang for their comments on an earlier version of this book chapter.

REFERENCES

[1] Bellovin, S. ICMP Traceback Messages. Oct. 2001. Internet Draft, Work in Progress. ⟨http://www.ietf.org/proceedings/02mar/I-D/draft-ietf-itrace-01.txt⟩.

[2] Brownlee, N., C. Mills, and G. Ruth. "Traffic Flow Measurement: Architecture." RFC 2722, IETF. Oct. 1999.

[3] Caceres, R. N. G. Duffield, A. Feldmann, J. Friedmann, A. Greenberg, R. Greer, T. Johnson, C. Kalmanek, B. Krishnamurthy, D. Lavelle, P.P. Mishra, K. K. Ramakrishnan, J. Rexford, F. True, and J. E. van der Merwe.

"Measurement and Analysis of IP Network Usage and Behaviour." *IEEE Commun. Mag.* **May** (2000): 144–151.

[4] Cao, J., D. Davis, S. Vander Wiel, and B. Yu. "Time-Varying Network Tomography: Router Link Data." *J. Am. Stat. Assoc.* **95** (2000): 1063–1075.

[5] Cisco Systems, Inc. "Cisco IOS Netflow." Aug. 2004. ⟨http://www.cisco.com/warp/public/732/Tech/nmp/netflow/index.shtml⟩.

[6] Cisco Systems, Inc. "Documentation." Aug. 2003. ⟨http://www.cisco.com/univercd/cc/td/doc/product/software/ios120/120newft/120limit/120s/120s11/12s_sanf.htm⟩.

[7] Cisco Systems, Inc. "New Feature Documentaion" Sept. 2003. ⟨http://www.cisco.com/univercd/cc/td/doc/product/software/ios120/120newft/120t/120t3/netflow.htm⟩.

[8] Clark, David D. "The Design Philosophy of the DARPA Internet Protocols." In *Proc. ACM SIGCOMM*, 106–114. New York: ACM Press, 1988.

[9] Duffield, N. G., and M. Grossglauser. "Trajectory Sampling for Direct Traffic Observation." *IEEE/ACM Transactions on Networking* **9(3)** (2001): 280–292.

[10] Duffield, Nick, ed. "A Framework for Passive Packet Selection and Reporting." Work in progress, Internet Draft draft-duffield-framework-papame-01, July 200004. ⟨http://psamp.ccrle.nec.de/⟩.

[11] Feamster, Nick, Jared Winick, and Jennifer Rexford. "A Model of BGP Routing for Network Engineering." In *Proceedings of the International Conference on Measurements and Modeling of Computer Systems (SIGMETRICS 2004)*, edited by E. G. Coffman Jr., Z. Liu, and A. Merchant. New York: ACM Press, 2004.

[12] Feldmann, Anja. "BLT: Bi-layer Tracing of HTTP and TCP/IP." In *Proc. World Wide Web Conference*, 321–335. May 2000. ⟨http://www9.org/w9cdrom/index.html⟩.

[13] Feldmann, A., A. Greenberg, C. Lund, N. Reingold, and J. Rexford. "NetScope: Traffic Engineering for IP Networks." *IEEE Network Mag.* **14(2)** (2000): 11–19.

[14] Feldmann, A., A. Greenberg, C. Lund, N. Reingold, J. Rexford, and F. True. "Deriving Traffic Demands for Operational IP Networks: Methodology and Experience." *IEE/ACM Transactions on Networking* **2(3)** (2000): 265–280.

[15] Floyd, S., and V. Jacobson. "Random Early Detection Gateways for Congestion Avoidance." *IEEE/ACM Transactions on Networking* **1(4)** (1993): 397–413.

[16] Floyd, S., and V. Jacobson. "The Synchronization of Periodic Routing Messages." *IEEE/ACM Transactions on Networking* **2(2)** (1994): 122–136.

[17] Floyd, S., and V. Paxson. "Difficulties in Simulating the Internet." *IEEE/ACM Transactions on Networking* **9(4)** (2001): 392–403.

[18] Fraleigh, C., C. Diot, S. Moon, P. Owezarski, D. Papagiannaki, and F. Tobagi. "Design and Deployment of a Passive Monitoring Infrastructure." In

Lectures Notes in Computer Science, vol. 2170, 556. New York: Springer, 2001. ⟨http://citeseer.ist.psu.edu/fraleigh01design.html⟩.

[19] Handelman, S., S. Stibler, N. Brownlee, and G. Ruth. "RTFM: New Attributes for Traffic Flow Measurement." RFC 2724, IETF. Aug. 2004.

[20] Hui, Joseph Y. "Resource Allocation for Broadband Networks." *IEEE J. Selected Areas in Communications* **6(9)** (1988): 1598–1608.

[21] Internet Protocol Flow Information eXport (IPFIX). Sept. 2003. ⟨http://net.doit.wisc.edu/ipfx/⟩.

[22] Jacobson,   Van,   C.   Leres,   and   S.   McCanne.   "Tcpdump." ⟨ftp://ftp.ee.lbl.gov/⟩.

[23] Keys, Ken, David Moore, Ryan Koga, Edouard Lagache, Michael Tesch, and K. Claffy. "The Architecture of the CoralReef Internet Traffic Monitoring Software Suite." In *PAM2001-A Workshop on Passive and Active Measurements*. RIPE NCC, 2001. ⟨http://www.caida.org/tools/measurement/coralreef/⟩.

[24] Labovitz, C., G. R. Malan, and F. Jahanian. "Internet Routing Instability." *IEEE/ACM Transactions on Networking* **6(5)** (1998): 515–528.

[25] Leland, Will E., Murad S. Taqqu, Walter Willinger, and Daniel V. Wilson. "On the Self-Similar Nature of Ethernet Traffic (Extended Version)." *IEEE/ACM Trans. on Networking* **2(1)** (1994): 1–15.

[26] McCanne, Steve, and Van Jacobson. "The BSD Packet Filter: A New Architecture for User-Level Packet Capture." In *Proc. Winter USENIX Technical Conference*. USENIX Assoc., 1993.

[27] Medina, A., N. Taft, K. Salamatian, S. Bhattacharyya, and C. Diot. "Traffic Matrix Estimation: Existing Techniques Compared and New Directions." In *Proc. ACM SIGCOMM, 2002*. New York: ACM Press, 2002.

[28] Paxson, Vern. "Bro: A System for Detecting Network Intruders in Realtime." *Computer Networks* **31(23/24)** (1999): 2435–2463.

[29] Perkins, David T. *RMON: Remote Monitoring of SNMP-Managed LANs*. Prentice Hall, 1998.

[30] Zhang, Yin, M. Roughan, N. Duffield, and A. Greenberg. "Fast, Accurate Computation of Large-Scale IP Traffic Matrices from Link Loads." *Proc. ACM SIGMETRICS, 2003*. New York: ACM Press, 2003.

[31] Savage, Stefan, David Wetherall, Anna Karlin, and Tom Anderson. "Practical Network Support for IP Traceback." *IEEE/ACM Transactions on Networking* **9(3)** (2001): 226–237.

[32] Snoeren, A. C., C. Partridge, L. A. Sanchez, Ch. E. Jones, F. Tchakountio, S. T. Kent, and W. T. Strayer. "Hash-Based IP Traceback." In *Proc. ACM SIGCOMM, 2001*. New York: ACM Press, 2001.

[33] Stallings, William. "Security Comes to SNMP: The New SNMPv3 Proposed Internet Standards." *Internet Protocol J.* **1(3)** (1998).

[34] Stallings, William. "SNMP and SNMPv2: The Infrastructure for Network Management." *IEEE Commun. Mag.* **36(3)** (1998): 37–43.

[35] Stallings, William. *SNMP, SNMP v2, SNMP v3, and RMON 1 and 2*, 3rd ed. Reading, MA: Addison-Wesley, 1999.

[36] Stewart, John W. *BGP4: Inter-Domain Routing in the Internet*. Reading, MA: Addison-Wesley, 1999.

[37] Tebaldi, C., and M. West. "Bayesian Inference on Network Traffic." *J. Am. Stat. Assoc.* **93(442)** (1998): 557–576.

[38] van der Merwe, J., R. Caceres, Y. Chu, and C. J. Sreenan. "mmdump: A Tool for Monitoring Internet Multimedia Traffic." *ACM Comp. Commun. Rev.* **30(5)** (2000): 48–59.

[39] Vardi, Y. "Network Tomography." *J. Am. Stat. Assoc.* **91** (1996): 365–377.

Internet Topology: Discovery and Policy Impact, Part I

Hongsuda Tangmunarunkit
Ramesh Govindan
Scott Shenker

We cannot begin to understand the large-scale structure of the Internet without having accurate representations of the Internet topology. Mercator is a program that uses hop-limited probes—the same primitive used in *traceroute*—to infer an Internet *map*. It uses *informed random address probing* to carefully explore the Internet Protocol (IP) address space when determining router adjacencies, it uses source-route capable routers wherever possible to enhance the fidelity of the resulting map, and it employs novel mechanisms for resolving *aliases* (interfaces belonging to the same router). This chapter describes the design of these heuristics and highlights the limitations in collecting information about Internet topology.

An accurate model of Internet routing paths also gives us a better understanding of the Internet's large-scale structure. Internet routing is determined by *policy*. In theory, policy can *inflate* shortest-router-hop paths. Using a simplified model of routing policy in the Internet, we obtain approximate indications of the impact of policy routing on

The Internet as a Large-Scale Complex System,
edited by Kihong Park and Walter Willinger, Oxford University Press.

Internet paths. Our findings suggest that routing policy does impact the length of Internet paths significantly. For instance, in our model of routing policy, some 20% of Internet paths are inflated by more than five router-level hops.

1   INTRODUCTION

The development of a router-level map of the Internet received little attention from the research community. This is perhaps not surprising given the perceived difficulty of obtaining a high-quality map using the minimal support that exists in the infrastructure. As we show in this chapter, it is useful, and possible, to get an approximate map of the Internet. Such a map is a first step toward trying to understand some of the macroscopic properties of the Internet's physical structure. Other potential uses of Internet maps have been described elsewhere [20] and we do not repeat them here.

This chapter documents a collection of heuristics, some well known and some obscure, for inferring the router-level map of the Internet. Of the many possible definitions of the word *map*, the one we choose for the purposes of this chapter is: a graph whose nodes represent routers in the Internet and whose links represent *adjacencies* between routers. Two routers are adjacent if one is exactly one Internet Protocol (IP)-level hop away from the other. In section 4.1, we discuss the implications of this definition, and describe how well our map collection heuristics allow us to infer the complete Internet map. Inferring the number, or IP addresses, of Internet hosts is an explicit non-goal.

Perhaps the only ubiquitously available primitive to infer router adjacencies is the *hop-limited probe*. Such a probe consists of a hop-limited IP packet,[1] and the corresponding Internet control message protocol (ICMP) response (if any) indicating the expiration of the IP *time-to-live (ttl)* field (or other error indicators). The *traceroute* tool uses this primitive to infer the path to a given destination. Generally speaking, all earlier mapping efforts [7, 9, 20] have computed router adjacencies from a sequence of traceroutes to different Internet destinations. The destinations to direct the traceroutes are usually derived from one or more *databases* (e.g., such as routing tables, the DNS, or a precomputed table of host addresses). Finally, with one exception, all earlier mapping efforts have attempted to map the Internet by sending hop-limited packets from a single location in the network. Section 2 describes in greater detail these efforts at mapping the Internet.

In this chapter, we describe the heuristics employed by our Internet mapper program, Mercator. Written entirely from scratch, Mercator *requires no input*;

---

[1]Internet protocol packets have an integer *ttl* field; this field is decremented by each intervening router. When the ttl field reaches zero, the router sends an ICMP message to the sender of the IP packet.

in other words, it does not use any external database in order to direct hop-limited probes. Instead, it uses a heuristic we call *informed random address probing*; the targets of our hop-limited probes are informed both by results from earlier probes, as well as by IP address allocation policies. Such a technique enables Mercator to be deployed anywhere because it makes no assumptions about the availability of external information to direct the probes. Mercator also uses *source-routing* to direct the hop-limited probes in directions other than radially from the sender. This enables Mercator to discover "cross-links"—router adjacencies that might otherwise not have been discovered. As we describe later, these heuristics can result in several *aliases* (interface IP addresses) for a single router. Mercator also contains some heuristics for resolving these aliases. Section 3 describes these heuristics in greater detail, and discusses the limitations of the resulting map.

In section 4, we discuss our experiences with a deployment of Mercator. We describe various techniques we have used to validate the map resulting from this deployment. We also present some preliminary estimates of the size of this map, and analyze some graph-theoretic properties of the map. Finally, section 5 summarizes our main contributions, and indicates directions for future work.

2 RELATED WORK

Several Internet mapping projects have attempted to obtain a router-level map of the Internet. The earliest attempt we know of [16] traced paths to 5000 destinations from a single network node. These destinations were obtained from a database of Internet hosts that, in 1995, had sent electronic mail to a particular organization. In addition, a small number (11) of these destinations were used as intermediate nodes in *source-routed* traceroutes to the remaining destinations. Although the use of source routing can result in greater map fidelity, it is unclear how complete the resulting map is, given that the chosen destinations essentially represent an arbitrary subset of hosts in the Internet.

More recently, researchers [7, 20] have used BGP backbone routing tables in order to determine the destinations of traceroutes. For each prefix in the table, they repeatedly generate a randomly chosen IP address from within that prefix. From traceroutes to each such address, they determine router adjacencies, building a router adjacency graph in this manner. As we show in section 3, this alone does not result in an Internet map as we have defined it. In particular, these techniques may miss backup links (for which Siamwalla et al. [20] proposes—but does not report results of—tracing from several locations in the Internet). Furthermore, these traceroutes may discover two or more interfaces belonging to the same router; these projects do not propose techniques for resolving such aliases. The *skitter* tool [9] uses a database of Web servers to determine traceroute targets. Finally, the *rocketfuel* tool uses similar techniques to infer the router-level topology of individual autonomous systems.

Techniques for collecting other representations of Internet, such as the autonomous system (AS) topology, have also been documented in the literature. In one approach, traces of backbone routing activity over a period of several days have been used to infer inter-AS "links" [14]. Each link represents an inter-ISP peering or a customer-ISP connection. Instantaneous dumps of backbone routing tables have also been used to infer AS-level links [5].

In some cases, router support can be used to determine router adjacencies. For example, Intermapper [10] builds a list of router adjacencies by recursively interrogating routers' SNMP [8] MIBs. A similar technique can also be used for the Internet's multicast overlay network, the MBone [21]. Routers on the MBone support an IGMP query that returns a list of neighbors.

Our interest in heuristics for Internet mapping are motivated by the desire to understand network structure better. In this chapter, we present some preliminary results from analyzing the resulting map. More generally, a high-quality Internet map can be used to validate compact topology models such as those proposed in Faloutsos et al. [12]. Such models, or the resulting Internet map itself, can be used as input to simulations [2]. Moreover, high-quality Internet maps can also be used to validate hypotheses about scaling limits of real networks [18].

Lately, a large body of literature has focused on measuring and characterizing Internet topology, and developing synthetic topology models. Broido and Claffy [6] discuss several properties of the router-level topology of the Internet, as measured by the skitter tool. While not much work has explicitly focused on synthetic models for router-level topologies, the work of Faloutsos [12] has sparked interest in topology models that produce graphs with power-law degree distributions. Two classes of such models have emerged. One class attempts to grow such graphs using elementary rules for connecting nodes. This class of topology models is inspired by, and exemplified by, the work of Barabasi et al. [3]. A second class of models [15] extracts node degrees from a power-law degree distribution and connects nodes heuristically or randomly in order to satisfy assigned node degree. Finally, Tangmunarunkit et al. [23] have shown that both these classes of topology generators produce models that match real graphs more closely than an earlier generation of topology models [25].

3   MAPPING HEURISTICS

In this section, we discuss the design of several heuristics for mapping the Internet. This design is driven by several goals and requirements, which we discuss first.

3.1   THE CHALLENGE

The challenge we set ourselves at the outset of this project was to find a collection of heuristics that would allow us to map the Internet:

- From a single, arbitrary, location, and
- Using only hop-limited probes.

We chose the first restriction for two reasons. First, deployment of the mapping software then becomes trivial, especially if the heuristics do not require a specific topological placement (e.g., at exchange points or within backbone infrastructures). In fact, the results described in this chapter were obtained by running the Mercator software on a workstation at the edge of a campus network. Second, while it is certainly feasible to design a distributed mapping scheme, we chose to defer the complexity of implementing such a scheme until after we had explored the centralized mapping solution. Upon first glance, requiring our mapping software to run from a single node might seem too restrictive. It would appear that the single perspective provided by this restriction could result in large inaccuracies in the map. As we show later in section 3.4, the use of source routing can help alleviate these inaccuracies.

The second restriction—using *only* hop-limited probes—makes only minimal assumptions about the availability of network functionality. This restriction follows from the first, allowing the mapping software to be deployed anywhere in the network. It also implies that we explicitly chose *not* to use any external databases (the DNS, routing table dumps) to drive map discovery. Not only does this choice pose an academically interesting question (how can we map an IP network starting with nearly zero initial information?), but it can also lead to more robust heuristics for map discovery. As we show later, not all backbones have routing table entries for all Internet addresses (sec. 3.5), and not all router interfaces are populated in the DNS (sec. 4.3).

There are several secondary requirements that inform the design of map discovery heuristics.

- Obviously, the resulting map must be *complete*. Clearly, this requirement conflicts with the single-location restriction; using hop-limited probes from a single location cannot possibly reveal all router adjacencies. In section 4.1, we argue that our heuristics can give us nearly complete maps of the *transit* portion of the Internet.
- The map discovery heuristics must not impose significant probing *overhead* on the network.
- Informally, the heuristics must not result in significantly slower map discovery compared to existing approaches. Rather than quantify this requirement, we chose to sacrifice rapidity of map discovery in favor of completeness and reduced overhead wherever necessary. This choice has interesting consequences, as described in section 4.1.

3.2   INFORMED RANDOM ADDRESS PROBING

Mercator uses hop-limited probes to infer router adjacencies, but does not use external databases to derive targets for these probes. In the absence of external information, one possible heuristic—with obvious convergence implications—is to infer adjacencies by probing paths to addresses randomly chosen from the entire IP address space [20]. In this section, we describe a different heuristic, *informed random address probing*.

The goal of this heuristic is to guess which portions of the entire IP address space contain addressable nodes. An addressable node is one which has IP-level connectivity to the public Internet. Because IP addresses are assigned in prefixes, Mercator makes *informed* guesses about which prefixes might contain addressable nodes. From within each such *addressable prefix*, it then uniformly randomly selects an IP address as the target for one or more hop-limited probes (described in sec. 3.3). Mercator uses two techniques to guess addressable prefixes:

1. Whenever it sees a response to a hop-limited probe from some IP address A, Mercator assumes that some prefix P of A must contain addressable nodes. In this sense, Mercator is *informed* by the map discovery process itself.
2. If P is an addressable prefix, Mercator guesses that the neighboring prefixes of P (e.g., 128.8/16 and 128.10/16 are the neighboring prefixes for 128.9/16) are also likely to be addressable.[2] This technique is based on the assumption that address registries delegate address spaces sequentially.

In the following paragraphs, we describe the details of informed random address probing.

In a given instance of Mercator, repeated application of these two techniques leads to a gradually increasing *prefix population*. In order for both the above techniques to work, the prefix population must be seeded with at least one prefix. Mercator uses the IP address of the host it is running on to infer this seed prefix. Then, technique 2 above ensures that this prefix population eventually covers the entire address space. Technique 1 attempts to ensure that addressable prefixes are explored early, leading to more rapid map discovery. Technique 1 alone is insufficient for complete map discovery. For some choices of seed prefix, using this technique alone might only result in a campus-local map.

For technique 1, we need a heuristic that infers the length of an addressable prefix P from a router's IP address A. In classful terms [19], if A is a class A (class B) address, Mercator assumes that P's prefix length is 8 (respectively 16). This heuristic is based on pre-CIDR [13] address allocation policies. If A is a class C adddress, Mercator assumes that P is a prefix of length 19. This assumption is based on the practice of some top-level address registries in allocating CIDR blocks of length not greater than 19. For simplicity, we chose relatively coarse-

---

[2]Note that Mercator does not rely on these neighboring prefixes being topologically related—e.g., assigned by the same ISP—to P.

grained guesses about the size of addressable prefixes. As a result, only a small portion of an addressable prefix P might actually contain addressable nodes. However, because we probe randomly (and repeatedly, as described in section 3.3) within P, the choice of the prefix length does not affect the completeness of the resulting map, only the rapidity of the discovery.

Finally, we need a heuristic that determines how often technique 2 is invoked, and which addressable prefix's neighbor is chosen. If, within some window T the application of technique 1 has not resulted in an increase in the prefix population, Mercator selects a neighbor of some existing prefix P. In our implementation, T is chosen to be 3 minutes. P is chosen from among those prefixes in the population at least one of whose neighbors is not in the population. Among these prefixes, P is selected by a lottery scheduling [24] algorithm. The number of lottery tickets for each prefix is proportional to the fraction of *successful probes* (sec. 3.3) on that prefix. This heuristic attempts to explore neighbors of those prefixes that are more densely populated with addressable nodes.

3.3 PATH PROBING

To discover the Internet map, Mercator repeatedly selects a prefix (in a manner described later in this subsection) from within its population, and probes the *path* to an address A selected uniformly from within that prefix. Like traceroute, Mercator sends UDP packets to A with successively increasing *ttls*. To minimize network traffic, the path probe is self-clocking—the next UDP packet is not sent until a response to the previous one has been received. The probing stops either when A is reached, or when a probe fails to elicit a response (within ten seconds), or a loop is detected in the path. Unlike traceroute, the latter two termination conditions are appropriate for Mercator since our interest is in inferring router adjacencies.

A path probe results in a sequence of routers $R_1, R_2, R_3, \ldots$ such that R_1 responded with an ICMP *time exceeded* for a UDP probe with ttl 1, and so on. From this sequence, Mercator inserts into its map nodes corresponding to R_1, R_2, etc., and links $R_1 \leftrightarrow R_2$, $R_2 \leftrightarrow R_3$, etc., if these nodes and links were not already in the map.

To reduce path probing overhead, not all path probes start at ttl 1. Rather, from the results of each path probe for prefix P, Mercator computes the furthest router R in that path that was already in the map at the time the probe was completed. Subsequent path probes to P start at the ttl corresponding to R. If the first response is from R, Mercator continues the path probe, otherwise it *backtracks* the path probe to ttl 1. This technique allows Mercator to avoid, where possible, rediscovering router adjacencies. Moreover, it reduces the probing overhead on routers in the vicinity of the host on which Mercator executes.

Rather than selecting prefixes in round-robin fashion, Mercator uses the lottery scheduling algorithm [24] where each prefix has lottery tickets proportional to the fraction of successful probes addressed to the prefix. A probe is deemed

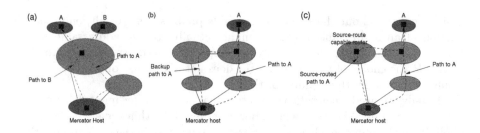

FIGURE 1   Continuous path probing does not only result in a shortest-path tree routed at the host running Mercator. The ellipses represent autonomous systems. Where available, source routing can lead to richer topology views.

successful if it discovers at least one previously unknown router. With lottery scheduling, then, Mercator's probing is biased towards recently created prefixes densely populated with addressable nodes. This heuristic attempts to speed up map discovery.

Finally, Mercator is designed to allow multiple path probes to proceed concurrently. This configurable number can be used as a tradeoff, giving up the rapidity of map discovery in order to reduce the overhead.

3.4   SOURCE-ROUTED PATH PROBING

Intuitively, in a shortest-path routed network, one might expect that path probing (sec. 3.3) results in a tree rooted at the host running Mercator. For two reasons, however, path probing actually discovers a richer view of the topology:

- Inter-domain routing in the Internet is policy-based [22]. Policies can result in widely divergent paths to two topologically contiguous routing domains (fig. 1(a)).
- Mercator continuously probes each addressable prefix over several days. It can, therefore, potentially discover *backup* paths to addressable prefixes (fig. 1(b)).

Even so, because Mercator attempts to discover the Internet topology from a single location, it may miss some "cross-links" in the Internet map. An obvious approach to increasing the likelihood of discovering all links in the Internet map is to run Mercator from different nodes in the network.

In this section, we describe a different solution to the problem, *source-routed path probing*. Essentially, our technique directs path probes to already discovered routers *via* source-route capable routers in the Internet (fig. 1(c)). Not all routers in the Internet are source-route[3] capable. In fact, only a small fraction (about 8%, see sec. 4) support source routing. In some cases, these routers are located at small ISPs who have neglected to disable this capability in their routers. However, a number of large backbone provider routers are also source-route capable. Presumably, these providers use source-routed traceroutes to diagnose connectivity problems in their infrastructures.

Can such a small number of source-route capable routers help us discover cross-links? Roughly speaking, each source-route capable router can give us the same perspective as an instance of Mercator running at the same location. How effective a small fraction of routers is in helping us discover the entire map depends on several factors including the structure of the Internet, the location of the host running Mercator, and the placement of these source-route capable nodes. To see whether there is even a *plausible* argument for the efficacy of source-routed path probing, we conducted the following experiment. We generated random graphs of different sizes with different average node degrees. For each such graph, we computed the percentage of links discovered as a function of the fraction of source-route capable nodes in the network. We used two methods for selecting source-route capable nodes: (a) nodes are equally likely to be source-route capable (the *uniform* method), and (b) source-route capable nodes are selected in decreasing order of degree (the *degree-based* method). This latter method is particularly relevant for *power-law* graphs (i.e., randomly [1] generated graphs with power-law degree distributions), which we also include in our experiment. We found (fig. 2) that for graphs larger than 1000 nodes, less than 1% of source-route capable routers suffice to discover over 99% of the links in the topology. This result holds for the random and power-law graph instances we tried, and for both the uniform and degree-based source-route capable router selection methods.

To detect source-route capability in a router R, we attempt to send a source-routed UDP packet *via* R to a randomly chosen router R_t from our map. This UDP packet is directed to a randomly chosen port and will usually elicit an ICMP `port unreachable` message from the target router. If we do receive such a response, we mark R source-route capable. We conduct this test for *every* router discovered by the path probing heuristic. Furthermore, for each router R, we repeatedly test for source-route capability (to limit overhead, only a small number of such tests are run concurrently). We do this because a given test may fail for reasons other than the lack of source-route capability: R may be unreachable at the time the test was conducted, R may not have a route to R_t, and so on.

Once a router R is deemed source-route capable, we repeatedly attempt to send path probes via R to randomly chosen *routers* from our map. This method of choosing probe targets is different from that used by path probing (sec. 3.3).

---

[3] IP packets can contain an optional *loose* source route. This is a sequence of IP addresses of intermediate network nodes that the packet is supposed to traverse before reaching its target. For various reasons, not all routers can process the source route option.

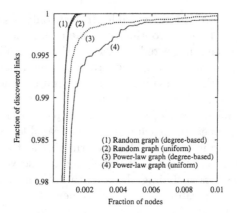

FIGURE 2   In large random and power-law networks, only a few source-route capable nodes ($<1\%$) are sufficient to discover 99% of the links.

We chose this approach only to avoid targeting hosts and setting off alarms—many host implementations automatically log source-routed packets. Even so, our probing activity was noticed by several system administrators (sec. 4.2). In order to reduce probing overhead, the source-routed path probe starts at R. The initial hop-count is inferred from the response obtained when R was first discovered. This initial hop-count guess might be incorrect (e.g., because the path to R has changed), but this does not affect the *correctness* of our map discovery. As before, we send hop-limited probes with successively increasing *ttls* terminating only if we reach the destination, a loop is detected in the path, or a small number (three, in our implementation) of consecutive hop probes have failed. This last criterion was empirically chosen to carefully balance the overhead of probing against the rapidity of map discovery.

3.5   ALIAS RESOLUTION

To simplify the preceding discussion, we said that path probes (source-routed or otherwise) discover *routers*. Path probes actually discover *router interfaces*. Thus, a single Mercator instance might discover more than one interface belonging to the same router (i.e., multiple *aliases* for the router). For example, because of policy differences, paths from the Mercator host to two different destinations can intersect (fig. 1(a)). Similarly, the primary and backup paths to a destination might overlap (fig. 1(b)). Finally, a source-routed path probe can, because it probes from a different perspective, discover additional router interfaces (fig. 1(c)).

Resolving these aliases is an important step for obtaining an accurate map. Unfortunately, these aliases cannot be resolved by examining the syntactic structure of interface addresses. This is because a router's interfaces may be numbered from entirely different IP prefixes; this commonly occurs at administrative boundaries. The DNS is equally ineffective for these purposes; where interfaces are assigned DNS names, different interfaces are assigned different names. At administrative boundaries, different interfaces of a router may actually be assigned names belonging to different DNS domains.

Our solution leverages a *suggested* (but not required) feature of IP implementations [4]. Suppose a host S addresses a UDP packet to interface A of a router (fig. 3(a)). Suppose further that that packet is addressed to a nonexistent port. In what follows, we call such a packet an *alias probe*. The corresponding ICMP *port unreachable* response to this packet will contain, as its source address, the *address of the outgoing interface for the unicast route towards S*. In figure 3(a), this interface is B. We have verified this behavior of IP stacks belonging to at least two major router vendors.

This feature suggests a simple heuristic for alias resolution. Send an alias probe to interface X. If the source address on the resulting ICMP message (assuming there is one) is Y, then X and Y are aliases for the same router. This heuristic was also suggested in earlier work [16]. However, Mercator uses two additional refinements necessary to correctly implement alias resolution.

The first refinement repeatedly sends alias probes to an interface address. To see why this is necessary, consider the following scenario. Suppose, as in figure 3(a), a router has two interfaces A and B. Suppose further that, at some instant, an alias probe to A returns A itself. This implies that, *at that instant*, the router's route to the sending host exits interface A. It is possible that, at some later instant, an alias probe to B returns B. This can happen if, at the instant the second probe is sent, the router's route to the sending host exits interface B. Mercator, therefore repeatedly sends alias probes to every known interface. To limit probing traffic, only a small number of alias probes are run concurrently. The efficacy of such repeated alias probing is based on the observation that there exist *dominant* paths [17] and routes [14] in the Internet. There is a high likelihood that, eventually, a large fraction of the alias probes will be returned via the router's dominant path to the sending host.

The second refinement uses source-routed alias probes. Figure 3(b) explains why this is necessary. In practice, large backbones do not have *complete* routing tables. Thus, the backbone(s) to which the sending host is eventually connected may not be able to forward an alias probe to its eventual destination. To circumvent this, Mercator attempts to send alias probes *via* source-route capable routers. This random, combinatorial search of all source-route capable routers is a time-consuming process. However, in the absence of other information about where one might find a route to a given interface, we believe this is the only option left to us.

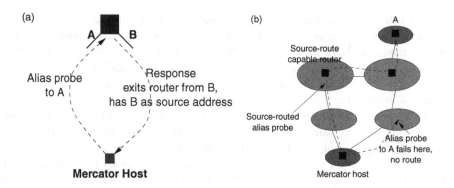

FIGURE 3 Repeated alias probing can resolve router aliases. In some cases, source-routed alias probes are necessary for alias resolution.

3.6 MERCATOR SOFTWARE DESIGN

Although some of the probing primitives that Mercator uses are available in the *traceroute* program, we chose to implement Mercator from scratch. Doing so allowed us the flexibility of trying different probing heuristics (e.g., different termination conditions for source-routed path probing, sec. 3.4). It also enabled us to carefully tune the overhead imposed by probing without sacrificing the rapidity of map discovery (e.g., the path probing optimization described in sec. 3.3).

Mercator is implemented on top of *Libserv*, a collection of C++ classes written by one of the authors that provides non-blocking access to file system and communication facilities. This allows Mercator to have several outstanding probe packets, and to independently tune the number of outstanding path probes, source-routed path probes, and alias probes. Together with Libserv, Mercator is about 14,000 lines of C++ code.

Mercator periodically checkpoints its map to disk. It is completely restartable from the latest (or any) checkpoint. We made several heuristic revisions during the implementation of Mercator, and this feature allowed us to avoid re-discovering sections of the Internet topology. It also allowed us to recover from buggy implementations of heuristics.

4 EXPERIENCES AND PRELIMINARY RESULTS

Although the heuristics described in section 3 seem plausible, several questions remain unanswered: Is the resulting Internet map complete? What techniques might we use to validate the resulting map? How well do the heuristics work? What does the map look like? This section attempts to answer some of these questions.

4.1   IMPLICATIONS OF OUR METHODOLOGY

A careful examination of the heuristics described in section 3 reveals several important implications of our methodology. The "map" discovered by Mercator is *not* a topology map—it does not enumerate all router interfaces, and does not depict shared media.

First, because it uses path probing, Mercator cannot discover all interface addresses belonging to a router. Instead, it discovers only those interfaces through which paths from the Mercator host "enter" the router. Source-routed path probing can help increase the number of interfaces discovered, but our use of source-routed path probing does not guarantee that all interfaces are discovered.

Second, Mercator does not implement heuristics for discovering shared media. To do this, it would have to infer the subnet mask assigned to router interfaces. Unfortunately, two potential probing techniques for inferring subnet masks do not work very well. The ICMP query that returns the subnet mask associated with an interface is not widely implemented. Sending ICMP echo requests to different broadcast addresses (corresponding to different subnet masks) and inferring the subnet mask from the broadcast request that elicits the most responses [20] does not work either—this capability is now disabled in most routers in response to *smurf* attacks. It may be possible, however, to infer shared media by isolating highly meshed sections of our Internet map; this remains future work.

Mercator's map is *not* an instantaneous map of the Internet. To reduce probing overhead, Mercator limits the number of outstanding path probes. As a result, it takes several weeks (sec. 4.2) to discover the map of the Internet. During this interval, of course, new nodes and links may have been added; Mercator is unable to distinguish these from pre-existing nodes. For this reason, the resulting map is a time-averaged picture of the network. Furthermore, Mercator discovers a time-averaged *routed* topology. If a router adjacency is used as a backup link, and that backup link is never traversed during a Mercator run, that adjacency will not be discovered. Similarly, if two routers are physical neighbors on a LAN, but never exchange traffic between themselves (e.g., for policy reasons), that adjacency is never discovered.

Finally, Mercator does not produce a *complete* map of the Internet. In particular, Mercator does not discover details of stub (campus) networks. Even though a single run of Mercator sends a large number of path probes, a given campus network is not probed frequently enough to discover the entire campus topology. As well, we believe that many campuses have source-routing turned off, so Mercator is unable to discover cross-links. However, we have a higher confidence in the degree of completeness of the Mercator map with respect to the *transit* portion of the Internet. This is because our probes traverse the transit portion more frequently (and this is true to a greater extent of the *core* of the Internet). We could obtain a more complete map by running multiple instances of Mercator and correlating the results. This is left for future work.

4.2 EXPERIENCES

We implemented the heuristics of section 3 and ran one instance of Mercator on a PC running Linux. When configured with a limit of 15 concurrent path probes, Mercator takes about three weeks to discover nearly 150,000 interfaces and nearly 200,000 links. These numbers are greater than the corresponding numbers obtained from Burch and Cheswick [7], serving as a simple validation of the completeness of our run. Before we discuss how we validated our Internet map, and what that map reveals in terms of the macroscopic Internet structure, we describe our early experiences in running Mercator and analyzing some of the data.

How well do our techniques for inferring addressable prefixes work? To answer this question, we first obtained the list of prefixes contained in a BGP table obtained from the Route Views archive. We then compared this with the prefixes inferred by Mercator. Only 8% of the prefixes in the routing table were not "covered" by any prefix inferred by Mercator. Conversely, 20% of prefixes inferred by Mercator did not have at least one overlapping prefix in the routing table. This latter figure is not surprising; our assumption that the "neighbor" of an addressable prefix is also addressable (sec. 3.2) can result in an overestimation of the addressable space. These numbers validate both the efficacy of the heuristic and the near completeness of our exploration of the address space.

How well does path probing work? There is a perception that many ISPs disable traceroute capability through their infrastructures. If this were true of the major US backbones, we would clearly not be able to see any routers belonging to these in our map. In fact, we do. There is also a perception that some ISPs configure their routers to not decrement *ttls* across their infrastructures. If this were true, our map would contain many routers with a large outdegree (i.e., the entire ISP appears to be one large router). We have found one possible instance of this, but no evidence that this practice is widespread. We conjecture that traceroute availability is higher than supposed by many researchers because ISPs use this capability to debug their infrastructures.

We also found some instances of obvious misconfiguration. Mercator received at least one ICMP *time exceeded* message whose source address was a Class D (multicast) address. It was also able to successfully probe the path to some net 10 addresses; such addresses are reserved for private use and should not be globally routable [13].

Our experience with alias resolution has been mixed. Mercator was able to resolve nearly 20,000 interfaces. However, many other interfaces were unreachable from the Mercator host. Mercator could not, therefore, determine whether such interfaces were aliases for already discovered routers. Three causes may be attributed to this:

- Some of these interfaces are numbered from private address space [13], and (with minor exceptions), this space is not globally routable. To these interfaces, then, the alias resolution procedure cannot be applied.
- Some others are assigned non-private addresses, but the corresponding ISP does not propagate routes to these interfaces beyond its border. To these interfaces, again, the alias resolution procedure cannot be applied.
- Finally, some of these interfaces are assigned non-private addresses, and some, but not all parts of the Internet core (and, in particular, the backbone ISP to which the Mercator host defaults) carry routes to these interfaces. To such addresses, we attempted source-routed alias resolution (sec. 3.5). This heuristic works, but is very slow, because it involves a random search among all source-route capable routers. As of this writing, we were able to resolve only about 3,000 interfaces using this technique.

To date, Mercator has discovered about 3,000 distinct[4] private addresses. A total of 15,000 interfaces are not directly reachable from Mercator, and it is unclear how many of these can be reached using source-routing.

Given the widely-held belief that "source-routing does not work," our experiences with source-routing path probing (sec. 3.4) are probably of most interest. We summarize our experiences below:

- About 8% (nearly 10,000) routers on the Internet *are* source-route capable. Of these routers, several belong to large U.S. backbones, implying that they are actively used (presumably to debug ISP infrastructures using source-routed traceroutes).
- Source-routed path probing works reasonably well (see the next bullet), although, for a given traceroute, not all intermediate hops respond. However, this is sufficient for our purposes, since our goal is to infer router adjacencies, rather than study paths. Nearly 1 in every 6 links in our map was discovered using source-routed probing alone (i.e., direct path probing did not subsequently rediscover these links).
- We did encounter two bugs in source routing. First, there is at least one router in the Internet that responds to source routed packets as though the packet was sent directly to it. That is, this router completely ignores the IP loose source route option. Second, some earlier versions of a router vendor's operating system nondeterministically fail to decrement the *ttl* on UDP packets with the loose source route packet. We do not know the extent of these bugs. Of these bugs, the latter has the potential for corrupting our map.

Finally, our experiences with the "social" consequences of sending traceroutes were similar to those reported elsewhere [7]. Many system administrators

---

[4]Private addresses, by definition, are not required to be globally unique, so more than one physical interface can be assigned the same physical address.

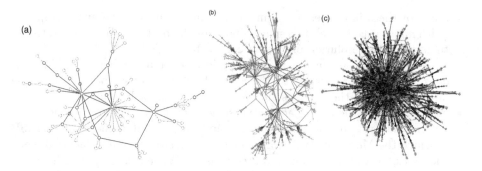

FIGURE 4 Mercator can be used to visualize, and analyze, ISP structures.

(largely from corporate sites containing firewalls which log UDP packets to non-existent ports) noticed our activity and registered abuse complaints with our institution. This is, in general, a practical problem associated with large-scale measurement experiments on an increasingly commercial Internet infrastructure. The conduct of these experiments has been simplified by adherence to two simple rules. First, mask measurement traffic so that it does not appear to be abnormal. For example, later runs of our Mercator tool switched from using UDP-based traceroutes to ICMP-based traceroutes. This helped reduce abuse complaints. Second, respond immediately and comprehensively to abuse complaints. Most system administrators are sympathetic to experimentation. Responding to their complaints also assures them that the measurement traffic is not an attempt to compromise their network.

4.3 VALIDATING THE MAP

How did we validate the resulting Internet map? The previous section discussed simple validation tests for our map. Could we have used the data from Burch and Cheswick [7] to carefully compare, on a link-by-link basis, our two maps? Although theoretically feasible, such a comparison would have required significant additional programming effort. That data set was gathered from a different location in the Internet. Consequently, it discovered different interface addresses than Mercator; to compare the two data sets, we would have had to resolve aliases (sec. 3.5) between the two sets of interface addresses. We have deferred such validation for future work.

Instead, we are currently validating *subgraphs* of our Internet map. Our strategy is to compare published ISP maps against ISP subgraphs extracted from our map. To extract ISP subgraphs, we infer routers belonging to an ISP from DNS names assigned to router interfaces (not all router addresses have corresponding names assigned in the DNS; we failed to complete the inverse mapping on nearly 30% of the discovered interfaces). Using this technique, and the *nam* network animator [11], we were able to visualize ISP structures (fig. 4). In some cases, ISPs use naming conventions that allow us to infer inter-city links, "core" routers, and customer connections. Using these, we can sometimes isolate the core portions of ISP networks; studying the structure of ISP networks is left for future work.

The extraction of ISPs is tedious enough, but the validation of the map is made much harder by the fact that most commercial ISPs do not publish router-level connectivity (this is regarded as proprietary information). To date, we have compared our discovered topology against the structures of a local ISP (Los Nettos) and a national backbone network (Verio). Mercator discovered all routers and all but one link in the local ISP. That these networks are topology close to the host which ran Mercator is probably irrelevant; our probing is not informed by topological closeness, so we expect that Mercator would show similar fidelity for distant, but similarly sized ISPs. For the backbone ISP, however, Mercator discovered all but 9 of the 62 nodes in the backbone; of the 82 links between the discovered nodes, Mercator discovered all but 16. These numbers are encouraging: they argue that from a single location, it is possible to achieve reasonably high map fidelity.

Overall, our validation results are somewhat encouraging. The technique appears to give a fairly complete picture of ISPs which might be considered to be at the edge of the Internet's transit portion. It gives us greater confidence in the hypothesis that Mercator can map the transit portion with reasonable fidelity. There is one caveat, however. The connectivity structures of both networks are relatively sparse. It is easy to see that Mercator can map such sparse ISPs quite well. It is less clear how successful Mercator will be in mapping more meshed ISP structures (e.g., those that consist of an ATM core with private virtual circuits established between routers). We are currently validating such structures in our map.

5 CONCLUSIONS AND FUTURE WORK

This chapter documents several techniques to infer the Internet map, and reports on our experiences in designing and experimenting with different heuristics for increasing the fidelity of the map. Clearly, our results have been mixed. We have been able to explore more of the Internet than previous efforts, but have, in the process, revealed several reasons why it is exceedingly difficult to obtain a highly accurate map in the existing Internet infrastructure.

We intend to explore several directions in the future; running Mercator from more than one location, evaluating heuristics for inferring topological elements such as shared media, and validating sections of the map with the help of ISPs in order to try to bound the degree of inaccuracy of our map.

ACKNOWLEDGMENTS

This work was done at the University of Southern California's Information Sciences Institute, 4676 Admiralty Way, Suite 1001, Marina del Rey, CA 90292. This work was partially supported by the Defense Advanced Research Projects Agency under grant DABT63-98-1-0007. Any opinions, findings, and conclusions or recommendations expressed in this material are those of the authors and do not necessarily reflect the views of the Defense Advanced Research Projects Agency.

REFERENCES

[1] Aiello, W., F. Chung, and L. Lu. "A Random Graph Model for Massive Graphs." In *Proc. of the 32nd Annual Symposium on Theory of Computing*, 171–180. New York: ACM Press, 2000

[2] Bajaj, S., L. Breslau, D. Estrin, K. Fall, S. Floyd, P. Haldar, M. Handley, A. Helmy, J. Heidemann, P. Huang, S. Kumar, S. McCanne, R. Rejaie, P. Sharma, K. Varadhan, Y. Xu, H. Yu, and D. Zappala. "Improving Simulation for Network Research." Technical Report 99-702, University of Southern California, Los Angeles, CA, 1999.

[3] Barabasi, A.-L., and R. Albert. "Emergence of Scaling in Random Networks." *Science* **286** (1999): 509–512.

[4] Braden, R. "Requirements for Internet Hosts—Communication Layers." Request for Comments 1122, RFC Index, Internet RFC/STD/FYI/BCP Archives, October, 1989.

[5] Braun, H.-W., and K. C. Claffy. "Global ISP Interconnectivity by AS Number." Measurement & Operations Analysis Team, NLANR. November 1977. ⟨http://moat.nlanr.net/AS/⟩.

[6] Broido, A., and K. C. Claffy. "Internet Topology: Local Properties." SPIE International Symposium on Convergence of IT and Communication. CAIDA, 2001. .

[7] Burch, H., and B. Cheswick. "Mapping the Internet." *IEEE Comp.* **32**(4) (1999): 97–98.

[8] Case, J. D., M. Fedor, M. Schoffstall, and C. Davin. "Simple Network Management Protocol (SNMP)." Request for Comments 1157, RFC Index, Internet RFC/STD/FYI/BCP Archives, May, 1990.

[9] Claffy, K. C., and D. McRobb. "Measurement and Visualization of Internet Connectivity and Performance." June 2001. ⟨http://www.caida.org/tools/measurement/skitter/⟩.

[10] Dartmouth University. "Intermapper: An Intranet Mapping and Snmp Monitoring Program for the Macintosh." 2000. ⟨http://www.dartmouth.edu/netsoftware/intermapper⟩.

[11] Estrin, D., M. Handley, J. Heidemann, S. McCanne, Y. Xu, and H. Yu. "Network Visualization with the VINT Network Animator Nam." Technical Report 99-703, University of Southern California, 1999.

[12] Faloutsos, C., M. Faloutsos, and P. Faloutsos. "What does Internet look like? Empirical Laws of the Internet Topology." In *Proceedings of ACM SIGCOMM 1999*, Boston, MA. New York: ACM Press, 1999.

[13] Fuller, V., T. Li, J. Yu, and K. Varadhan. "Classless Inter-Domain Routing (CIDR): An Address Assignment and Aggregation Strategy." Request for Comments 1519, RFC Index, Internet RFC/STD/FYI/BCP Archives, September, 1993.

[14] Govindan, R., and A. Reddy. "An Analysis of Internet Inter-Domain Topology and Route Stability." In *Proceedings of INFOCOM '97: Sixteenth Annual Joint Conference of the IEEE Computer and Communications Societies*, 850–857. New York: IEEE Press, 1997.

[15] Jin, C., Q. Chen, and S. Jamin. "Inet: Internet Topology Generator." Technical Report CSE-TR-433-00, EECS Department, University of Michigan, Ann Arbor, MI, 2000.

[16] Pansiot, J.-J., and D. Grad. "On Routes and Multicast Trees in the Internet." *ACM SIGCOMM Comp. Comm. Rev.* **28(1)** (1998): 41–50.

[17] Paxson, V. "End-to-End Routing Behavior in the Internet." In *Proceedings of the ACM SIGCOMM Conference on Applications, Technologies, Architectures, and Protocols for Computer Communications*, 25–38. New York: ACM Press, 1996.

[18] Phillips, G., H. Tangmunarunkit, and S. Shenker. "Scaling of Multicast Trees: Comments on the Chuang-Sirbu Scaling Law." In *Proceedings of the conference on Applications, Technologies, Architectures, and Protocols for Computer Communication (SIGCOMM '99)*, 141–151. New York: ACM Press, 1999.

[19] Postel, J. "Internet Protocol." Request for Comments 791, RFC Index, Internet RFC/STD/FYI/BCP Archives, September, 1981.

[20] Siamwalla, R., R. Sharma, and S. Keshav. "Discovering Internet Topology." July 1998. ⟨http://www.cs.cornell.edu/skeshav/default.htm⟩.

[21] Reddy, A., D. Estrin, and R. Govindan. "Large-Scale Fault Isolation." Technical Report 99-706, Computer Science Department, University of Southern California, Los Angeles, CA, 1999.

[22] Rekhter, Y., and T. Li. "A Border Gateway Protocol 4 (BGP-4)." Request for Comments 1771, RFC Index, Internet RFC/STD/FYI/BCP Archives, March, 1995.

[23] Tangmunarunkit, H., R. Govindan, S. Jamin, S. Shenker, and W. Willinger. "Network Topology Generators: Degree-Based vs. Structural." In *Proceedings of ACM SIGCOMM '02*. New York: ACM Press, 2002..

[24] Waldspurger, C. A., and W. E. Weihl. "Lottery Scheduling: Flexible Proportional-Share Resource." In *First Symposium on Operating Systems Design and Implementation (OSDI)*, 1–11. USENIX Association, 1995.

[25] Zegura, E., K. L. Calvert, and M. J. Donahoo. "A Quantitative Comparison of Graph-Based Models for Internet Topology." *IEEE/ACM Trans. in Networking* **5** (1997): 6.

Internet Topology:
Discovery and Policy Impact, Part II

Hongsuda Tangmunarunkit
Ramesh Govindan
Scott Shenker

In this chapter, we present our initial findings of the impact of policy on Internet paths. In particular, our findings reveal answers to the following questions: How does policy based routing inflate Internet paths? For a source-destination pair, does there exist a detour path and how good is the best detour path compared to the policy path? Does policy routing funnel Internet paths through larger autonomous systems (ASs)?

1   INTRODUCTION

The earliest internet routing protocols attempted to construct the lowest delay paths to destinations [10]. Thereafter, based on operational experience with the stability of delay-sensitive routing [8], deployed routing protocols evolved to essentially supporting global shortest hop-count routing [4].

Today's Internet contains several administrative domains (or Autonomous Systems, ASs). Within a domain, routing computes the shortest weighted paths, based on administratively configured weights. Routing between domains is determined by *policy*. Each autonomous system (AS) can, based on configured policy, independently select routing information from its neighboring ASs, and selectively propagate this information. These policies are not expressed in terms of hop-distance to destinations. Depending on how these policies are constructed, then, the resulting policy-based paths to destinations may incur more router-level hops than shortest-router-hop path routing.

Here, we ask the question: By how much does this hierarchical form of routing affect Internet paths? This question was motivated by recent work [14] that observed that, for a significant fraction of Internet paths, there existed an intermediate node such that the composite path through the intermediate exhibited better performance (delay, throughput). In other words, routing in the Internet does not result in delay- or throughput-optimal paths. Perhaps this anomaly can be rectified by changing the Internet's routing infrastructure to be delay or load sensitive. Before we do this, however, it would be appropriate to understand how much of these observations can be explained by the fact that routing hierarchy and policy can result in longer hop paths. Our work takes the first step toward this goal. Understanding this question can also be important for understanding the overall efficiency of the Internet's routing infrastructure. Finally, an answer to this question can also inform protocol evaluation studies, which typically assume shortest-router-hop path (henceforth, shortest path) routing.

To understand how policy routing affects Internet paths, we use a simplified model of inter-AS routing policy that we call *shortest AS path*[1] (sec. 2). Even though, in theory, routing policy can be completely arbitrary, many—but not all—existing routing policies are based on shortest AS paths. To infer the router-level path corresponding to this policy, we first begin with a router-level map of the Internet. On this map, we assign routers to ASs and obtain an *AS overlay* on top of our router-level map. This construction enables us to compare the router-level *policy path* between any two nodes with the shortest path on the map. Each of the steps in our construction represents a simplification of reality. As such, then, our results are only approximate indications of the impact of policy on Internet paths. However, at each stage, we carefully validate our construction using a collection of actual traceroutes that represent real paths generated by policy-based routing. This gives us some confidence that our conclusions are meaningful.

We find several surprising results (sec. 3). On average, about 20% of Internet paths are inflated by 50% or more. We also find that about half of the source-destination pairs benefit from a *detour*. For these pairs, there exists an intermediate node—a detour—such that the overall policy path length through this intermediate node is less than the policy path between source and destination.

---

[1] In section 4.2, we study the sensitivity of our conclusions to the choice of a slightly more sophisticated routing model.

To our knowledge, no related work has addressed the impact of routing policy on Internet paths. Our work, however, complements many pieces of recent work aimed at understanding the structure of the Internet, and the properties of its paths.

- Several efforts have focused on discovering router level topologies of the Internet [1, 3, 7]. Such mapping efforts are the crucial first step to helping us understand the impact of routing policy.
- Other work has empirically studied the availability characteristics of paths [11], the loss and packet delivery performance of paths [12], and the existence of alternate paths with lower delay or higher throughput [14]. By considering hop-distance between network nodes, our work examines the potential inefficiencies resulting from policy routing.
- Finally, more recent work has looked at macroscopic properties of the inter-AS topology [5, 13]. By relating the AS structure to the underlying router-level map, our *AS overlay* may be able to explain some of the observed macroscopic properties of the inter-AS topology in terms of the underlying physical structure.

To place our work in context, we point out two important caveats. First, it is well known that hierarchical routing can result in non-optimal paths [9]. Our work quantifies the extent to which hierarchical routing in the Internet, together with routing policy, affects paths. Second, the correlation, if any, between path length and end-to-end delay is poorly understood. As such, then, our results cannot be directly extrapolated to observed delays on Internet paths. Nevertheless, our results are interesting, since path hop-count is the yardstick by which today's operational routing protocols are measured. Whether hop-count is the right metric for routing protocols is an orthogonal question that we do not address.

2 METHODOLOGY

Our first step in understanding the impact of routing policy on Internet paths was to obtain an Internet map. We used *Mercator* [15, sec. 3] for this purpose. In brief, Mercator uses hop-limited probes—the same technique used in traceroute—to infer a router-level Internet map. It employs several heuristics to obtain as complete a representation of Internet topology as possible. One heuristic, informed random address probing, carefully explores the IP address space for addressable routers and hosts. Mercator also exploits source-route capable routers wherever possible to help discover cross-links and thereby enhances the fidelity of the resulting map. Finally, it implements a technique for resolving interface aliases (interfaces belonging to the same router).

The map we used was collected between March 26, 2000, and April 10, 2000. Our map has 102,639 nodes and 142,303 links. This map is smaller than the maps reported in Govindan and Tangmunarunkit [7]. Despite this, we believe, as Govindan and Tangmunarunkit [7] do, that we have captured the transit portion of the Internet core, where policy impacts paths. In addition to the Internet map, we also collected the 61,485 traceroutes used in inferring the map. Because these traceroutes represent actual policy paths, we were able to use them to validate our policy model (see below).

Next, we attempted to compute an *AS overlay* on top of this Internet map. To do so, we assigned an autonomous system (AS) number to each router in our Internet map. For this, we used a BGP routing table from the Route Views archive. For a given router interface address, we found the matching route entry in the BGP routing table. We then assigned that router to the origin AS in the AS path associated with the route entry. This technique works for globally routed addresses; Tangmunarunkit et al. [15, sec. 3] describes situations where some ASs number their routers from private address space. These addresses represent potential inaccuracies in our AS overlay computation. There were 3210 such private interface addresses in our map; to these we assigned a designated unused AS number.

Not all IP addresses had matching entries in the BGP routing table. Furthermore, in the routing table, route aggregation can mask the actual origin AS. For these reasons, we also used the RADB (whois.radb.net) to determine the origin AS. Finally, if a router had many interfaces corresponding to different AS numbers, we picked the most frequently assigned AS for that router. In spite of using these two sources of information, we were unable to resolve the AS numbers for 497 non-private IP addresses. To these we assigned a designated unused AS number.

After each node was assigned an AS number, we then applied a simple collapsing algorithm to generate the AS overlay. The collapsing algorithm recursively marks neighbors belonging to the same AS with the same "color." Each color represents a node in the AS-level map. However, due to our incomplete information (both in the Internet map [7] and in AS number assignment), there were many disjoint clusters of nodes belonging to the same AS. In most cases, we found that such ASs normally have one large component with many small components, each with a small number of nodes. We solved this problem by identifying the biggest cluster of each AS and reassigning the smaller components (about 20,000 nodes total) to the topologically nearest AS.

Clearly, there are several sources of error in both the collected map, and in our AS overlay generation technique. The best we could have hoped for is that our techniques result in a first-order approximation of the actual inter-AS overlay. To verify this, we compared some macroscopic properties (such as the degree distribution, degree rank distribution, and hop-pair distribution[2]) of the resulting AS overlay to those of an AS-level topology inferred from a BGP routing table collected on April 10, 2000 [5] (fig. 1). Notice that these macroscopic properties are in qualitative agreement.

As an additional validation, we compared the collection of AS paths in a BGP routing table dump with the shortest AS path between the two corresponding nodes in our AS overlay. Figure 2(a) shows the distribution of the path length difference between node pairs in our map and the routing table dump. About 95% of node pairs in our overlay are within one AS hop of the corresponding path in the BGP routing table.

The final step in our methodology was to select a plausible model for Internet routing policy. In general, there exist two kinds of inter-ISP relationships, a provider-customer relationship and a peering relationship. To our knowledge, the prevalent and widespread routing policy practice is that an ISP picks the shortest AS path-based route for customer and peer supplied routes. Furthermore, an ISP only propagates its customer supplied routes to its peers, never routes supplied by other peers. This latter rule can result in AS-level paths not corresponding to the shortest AS path. In the absence of information about the exact nature of inter-ISP relationships, we use a simplified model in which the policy path between two ASs is determined by the shortest AS path between them. As Gao [6] has recently shown, it is possible to infer these relationships by simple examination of AS paths in BGP routing tables. We intend to refine our policy model based on this approach.

Clearly, this is only an approximate model of routing policy. We validate this model by comparing the length of the AS path corresponding to each traceroute in our collection, with the length of the shortest AS path between the corresponding source and the destination. Figure 2(b) shows the path-length comparison between the policy paths and the shortest AS paths of the collected traceroutes. We found that the shortest AS path underestimates the traceroute AS path by one AS hop or less for 70% of the traces, and less than two hops for 95% of the traces. Though the difference seems small, it doesn't represent spectacular agreement with our model, since many of the AS paths are relatively small (five hops or so). Nevertheless, this validation (and another described in sec. 3.1) is good enough to encourage us to pursue our initial understanding of the impact of routing policy on Internet paths.

3 RESULTS

Having a model for AS policy, we can now compute, for any given pair of nodes in the Internet map, the router-level path between them as determined by routing

---

[2]The degree rank distribution plots the degrees of nodes in the graph, sorted in decreasing order of degree. The hop-pair distribution plots the number of node pairs separated by a certain number of hops.

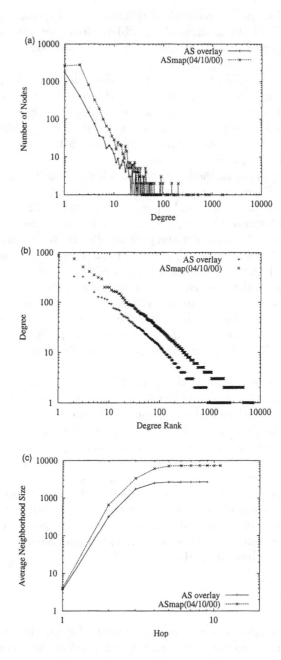

FIGURE 1   Macroscopic comparison of AS overlay and the AS topology.

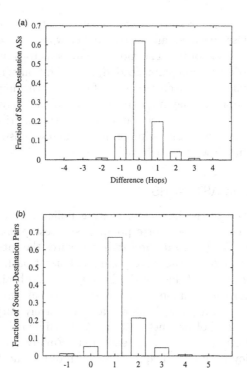

FIGURE 2 Validating the AS overlay and the shortest AS path model.

policy. We call such a path the *policy path*. Furthermore, we can also compute the router-level *shortest path* between the same pair of nodes.

Being thus empowered, we analyze the impact of routing policy on Internet paths by asking the following four different questions. These questions are all complementary ways of looking at how routing policy skews paths in the Internet.

- By how much does policy *inflate* unicast paths? That is, how different is a policy path from the corresponding shortest path?
- For nodes A and B in the Internet map, does there exist an intermediate node I such that the sum of the policy paths from A to I and from I to B is less than the policy path from A to B? If such an intermediate exists, A and B can circumvent routing and communicate using fewer hops via I.
- Does policy routing *funnel* Internet paths through larger ASs?
- How does policy routing impact multicast tree sizes?

Before we discuss our results, a couple of caveats are worth mentioning. First, unless otherwise specified, all our results below are derived from our policy model, not from the traceroutes. The model enables us to compute policy paths for arbitrary source destination pairs. Second, our definition of an "Internet path" implicitly considers paths between each pair of nodes in the map equally likely. In practice, for example, there may not exist any end-to-end conversations between two backbone routers. To more accurately model the distribution of Internet paths, it is necessary to have an estimate of how many "hosts" are attached to each router, an estimate we do not have.

3.1   INFLATING UNICAST PATHS

Our first examination of the impact of routing policy considers the difference between policy paths and their corresponding shortest paths. Specifically, we look at two metrics. The first, *inflation ratio* measures the ratio of the length of a policy path to the length of the corresponding shortest path. Figure 3(a) plots the cumulative distribution of this metric. This figure was obtained by computing the inflation ratio for each random source-destination pair in our Internet map, a total of 100,000 random pairs. It shows a quantatively surprising impact of routing policy. Some 20% of Internet paths are inflated by more than 50%. Some policy paths are inflated by a factor of nearly four. Finally, for only a fifth of the paths does the policy path length equal the length of the shortest path.

A second metric, the *inflation difference*, provides an alternative view of the impact of routing policy. This metric represents the absolute difference, in terms of the number of router hops, between the policy path and the shortest path. Figure 3(b) plots the cumulative distribution of inflation difference. For nearly 20% of the node pairs, the policy path is longer than the shortest path by more than 5 hops. Furthermore, there exist some node pairs for which the policy path is longer by 25 hops than the shortest path.

Perhaps a more reasonable way to consider the impact of routing policy might be to evaluate inflation for node pairs whose shortest paths are of the same length. Thus, figure 4(a) plots the cumulative distribution of the inflation ratio for four different shortest path lengths. Not surprisingly, in a distribution sense, longer paths are less inflated in proportion to their lengths—in general, shorter paths have more "room" for inflation than longer paths. Furthermore, the smaller lengths (notably length five) have significantly long tails.

Finally, figure 4(b) depicts the inflation difference for different shortest path lengths. This shows an interesting trend, namely that longer paths are more absolutely inflated than shorter paths. The exception to this observation is the length 20. We conjecture that the explanation for this exception is that really long paths have less absolute "room" to "grow." This observation has interesting consequences that we explore in section 3.2.

A subtle point that underlies the results presented in this and the next sections is that, for a given node pair, there may exist many different "shortest"

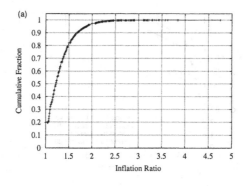

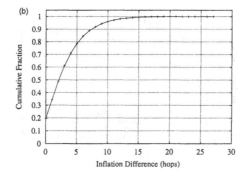

FIGURE 3   Cumulative distribution of inflation.

AS paths. The corresponding router-level policy paths for each possible shortest AS path may be widely different. For example, if shortest AS path X traverses larger ASs than shortest AS path Y, one might expect that the policy path corresponding to X might be longer than that for Y. Because it is computationally difficult to enumerate all possible shortest AS paths, all results presented here were derived by essentially randomly selecting shortest AS paths.

   Does this discrepancy affect our conclusions? One way to answer this question is to compare the inflation ratio and difference distributions for our collection of traceroutes (sec. 2) with the distributions presented above. Figure 5 does this. Our shortest AS path computation yields more conservative inflation ratios, and similar inflation differences, to the traceroute data. We also tried other approaches to selecting shortest AS paths, and found that most of the results presented in section are relatively insensitive to the way we pick the shortest AS path. Finally, we also compared our inflation ratios and differences to those

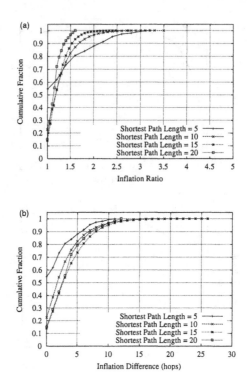

FIGURE 4   Cumulative distribution of inflation by path length.

obtained from a different mapping effort [2].[3] Figure 6 shows that our policy
model results in fairly conservative inflation ratios and differences.

3.2   FINDING DETOURS

We have seen that longer policy paths are generally more inflated in the number
of hops than shorter ones. This gap motivates the questions we examine in this
section: for what fraction of node pairs do there exist better alternate paths
(called *detours*) and by how much are these detours better than the policy paths?

Before we answer these questions, we more carefully define a detour. Consider
a pair of nodes A and B. We say that there exists a detour between A and B if
there exists an intermediate node I such that:

---

[3]To measure the inflation metrics on this collection of traceroutes, we need to create an
instance of Internet map corresponding to the traceroute collection. For this, we applied the
alias resolution technique described in Govindan and Tangmunarunkit [7], then synthesized
the Internet map.

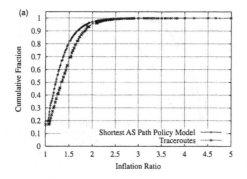

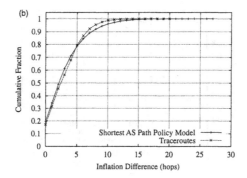

FIGURE 5   Comparison of inflation ratio and difference for traceroutes and for shortest AS path.

- I lies in a different AS than A or B,
- the AS path from A to I and from I to B is collectively *longer* than the shortest AS path between A and B, and
- the sum of the router-level policy paths between A and I and between I and B is less than the policy path between A and B.

Intuitively, a detour represents a way to circumvent routing by relaying communication between A and B at the application level.

To answer the above questions, we randomly generate a large number of source-destination pairs in which source and destination are selected from different ASs. For each source-destination pair, we find a policy path and the best detour path (if there exists one), then measure the difference between the two paths. We define two metrics for quantifying detours. The detour *gain* is the absolute difference, in router-level hops between the policy path and the best detour path. We define the gain to be zero if there exists no detour for that

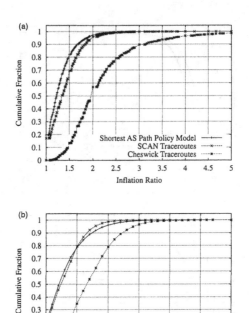

FIGURE 6   Comparison of inflation ratio and difference for Lucent traceroutes and for shortest AS path.

particular policy path. The detour *gain ratio* is the ratio between the gain and the length of the policy path between source and destination.

Figure 7 shows the cumulative distribution of the two metrics. Surprisingly, for 50% of the sampled paths, there exist superior detour paths. Furthermore, 20% of the sampled paths have a detour that is 3 hops or more less than the corresponding policy path. An alternative result is that 20% of the paths have a detour that is more than 20% shorter. Our finding is in quantitative agreement with that of Savage et al. [14], performance whilst ours is based only on path length. To what extent our finding is an explanation for theirs is a little unclear.

3.3   PATH CONCENTRATION AROUND LARGE AUTONOMOUS SYSTEMS

The previous two sections have studied the impact of routing policy on Internet paths. In this section, we look at a slightly different question: does routing policy force Internet paths through larger ASs? This question is one aspect of a larger, more general, question: Is topological connectivity rich enough that the logical

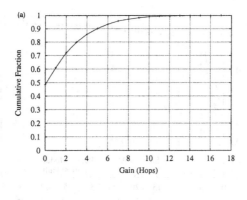

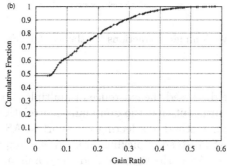

FIGURE 7   Detour gain and gain ratio distributions.

connectivity imposed by policy routing significantly skews the paths Internet traffic takes?

To study this question, we define for each node pair the *dominant AS* to be the largest (by size) transit AS encountered in the path between the two nodes. For each AS, we then define a *dominance fraction*; the fraction of node pairs for which that AS is the dominant one. We are interested in the correlation between size and the dominance fraction. In computing our dominance fraction, we do not consider node pairs which lie within a single AS, or in adjoining ASs. These, by definition, do not have any transit ASs between them.

We measure this metric both for our policy approximation, as well as for shortest router-level paths. Figure 8 shows that, regardless of whether policy routing is used, or shortest paths are used, the top 15 ASs by size dominate about 90% of the paths. This number is surprising not only in its magnitude, but also in the absence of any qualitative difference in dominance correlation between policy routing and shortest path routing.

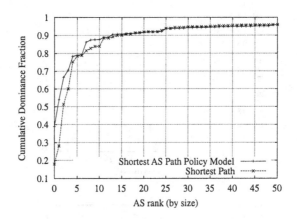

FIGURE 8   Cumulative dominance fractions for the top 50 ASs.

In section 3.1, we discussed the existence of multiple shortest AS paths. Unlike other results, it turns out that the dominance correlation is at least quantitatively affected by our method of selecting the shortest AS path. In particular, if we choose smaller ASs instead of picking larger ASs to explore in our breadth-first search for shortest AS paths then we find the dominance correlation curve to be different. Even by pessimistically picking the shortest AS path, we find that almost 70% of the paths are dominated by one of the top 25 ASs. We also checked to determine whether our results were sensitive to the method of selecting router-level shortest paths. They were not.

3.4   MULTICAST

A final question we look at is the effect of policy on multicast tree sizes. For this, we assume a simple random source and receiver placement, and compute source-rooted multicast trees using shortest paths and policy paths. We then compare the relative sizes of the shortest-path tree and the policy tree.

Figure 9 shows the ratio between the size of a policy tree and the corresponding shortest-path tree as a function of the number of receivers. In the same figure, we also show the average ratio between the unicast path length of a policy path versus a shortest path from the source to all receivers. As expected, the average unicast ratio is independent of the number of receivers in the group. This number is consistent with the average inflation ratio shown in figure 3(a).

However, for a small number of receivers, the policy trees are actually larger by about 30% than shortest-path trees. As the number of receivers increases, policy trees continue to grow larger than shortest-path trees. This result is somewhat counter-intuitive, since one might expect more path sharing with policy routing

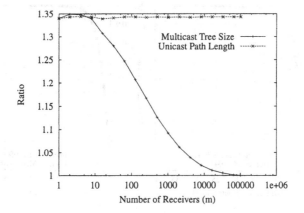

FIGURE 9   The effect of policy on multicast tree sizes

between receivers in the same AS, and hence smaller policy trees. However, the reduction of path sharing is probably offset by the overall increase in path length due to inflation.

4   SENSITIVITY OF OUR RESULTS

How sensitive are our results to (a) the particular snapshot of the Internet we use, and (b) the assumption that the shortest AS path is a reasonable representation of Internet routing policy? In this section, we reexamine our results from section 3 using a different Internet topology snapshot, and using a more refined routing policy model that does not violate peering arrangements. We find that our prior observations regarding the path inflation due to routing policy appear to hold both across time and with respect to a more sophisticated model of routing policy.

4.1   USING A DIFFERENT SNAPSHOT

To study the degree of Internet path inflation across time, first we needed to collect another snapshot of the Internet topology. We did this by running a slightly more optimized version of Mercator [15, sec. 1]. The map was collected between May 01, 2001, to May 06, 2001, approximately a year after the previous map (sec. 2) was collected. It contains 170,589 nodes and 215,385 links. On this map, we computed the AS overlay as described in section 2, and verified that its large-scale properties were qualitatively consistent with that of the AS map obtained from BGP routing table dumps.

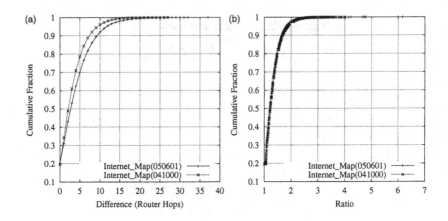

FIGURE 10   Inflation difference and inflation ratio.

Figure 10 shows the cumulative distribution of the paths with respect to the inflation difference and ratio, both for our earlier topology snapshot, as well as our more recent snapshot. We observe that the Internet path inflation with respect to the older map, when compared to the newer map, is more conservative according to the inflation difference and is approximately the same according to the inflation ratio. Moreover, we also find that the average path length on the newer map is longer than the average path length on the older map. This is possibly due to the bigger size of the newer map, in other words, the new map is about 50% larger than the older one.[4] Therefore, it is not so surprising that paths in the newer map reveal a greater inflation difference than those in the older map. Nevertheless, inflation has not changed *qualitatively* between the two snapshots.

Finally, in section 3.1 we also observed that the longer paths, when compared to the shorter ones, are more inflated in terms of the absolute difference but are less inflated in proportion to their lengths. We notice the similar behavior in the more recent snapshot as well (fig. 11). Thus, these findings seem to suggest that our observations regarding the Internet path inflation due to policy may hold across different snapshots of the Internet separated in time.

4.2 USING MORE REALISTIC POLICY

Our shortest AS path policy model suffers from one important drawback. Although the model enables us to study router-level paths between any two nodes in the network, it doesn't take into consideration the peering relationships among

---

[4]Whether this is the result of the growth in the Internet or a better router discovery technique revealing more information remains an open question.

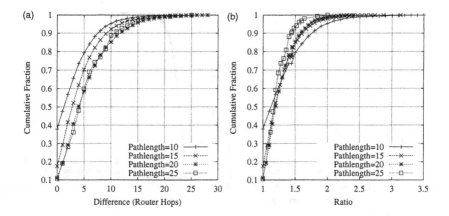

FIGURE 11 Inflation difference and inflation ratio with respect to different shortest-path-lengths.

AS nodes. As a result, some of the generated paths may not be realistic in the Internet context, (i.e., they may violate peering relationships by transiting through a stub domain in between two transit-domains). As an example, an AS path traversing through MCI-USC-SPRINT is an invalid path since USC is a customer of MCI and the packets between the national ISPs should never transit through one of their customers.

Fortunately, recent work by Gao [6] has described a more realistic technique for inferring AS peering relationships, e.g., provider-customer, peer-peer, or sibling-sibling relationship. Their work makes two assumptions: (1) that there is a strong correlation between the AS degree and AS size, for example, an AS with larger degree (i.e., an AS with many connectivities to its peers) is a bigger AS domain in size, and (2) that the AS paths are hierarchical. They assume that one signature of a hierarchy is that paths may go up, down, or up and then down the hierarchy. A path connecting two regional ISPs must traverse up the hierarchy to the national ISP, then the two national ISPs exchange packets and packets go down the hierarchy to the destination ISP. They apply the two assumptions to classify the types of AS paths or routes that can appear in BGP routing tables and then infer the peering relationship based on this classification. Using their technique, we annotate our AS overlay map with peering relationships. We then improve our policy model to consider only valid shortest AS paths, namely those that do not violate peering agreements.

We plot cumulative distribution of inflation difference and inflation ratio with respect to the modified routing policy model[5] in figure 12. For comparison,

---

[5]There are about 6.7% of sampled node pairs that are not reachable. We ignore these pairs in our inflation distribution.

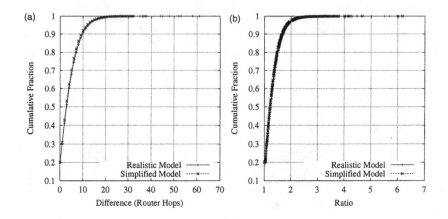

FIGURE 12   Inflation difference and inflation ratio by the realistic and simplified routing policy model.

we also include the earlier plots of inflation difference and inflation ratio of paths with respect to the simplified policy model in the plot. Our results indicate that the degree of inflation with respect to the two models—the simplified model and the realistic model—are qualitatively similar. Thus, *for determining the extent of inflation*, a shortest AS path model appears to suffice.

5   CONCLUSIONS

Does policy have an impact on Internet path length? This chapter clearly makes the case that it does, even with realistic models of routing policy. In our model, nearly 50% of paths benefit from a detour. Some small multicast trees are enlarged almost 30% by policy.

While our shortest AS path approximation may be rendered obsolete by more complicated routing policy, there exists a more enduring representation of our work. Shortest AS path represents the routing that would have resulted from a pure (policy-free) hierarchical routing in the Internet. In this sense, this work quantifies the impact on Internet paths of the particular instance of hierarchy that the Internet has evolved to today.

The eventual goal of topology discovery and policy modeling is the development of accurate models for Internet topology and routing. Such models can be used to understand the scaling of protocol behavior, the efficacy of attack prevention mechanisms and so on. Much work needs to be done to achieve this goal. Structural models have most promise—these models contain elemental mechanisms that mirror the topology construction and routing exchange in the real-world. Autonomous sytems are likely to be building blocks in such models. Thus, understanding AS topology and the role of geography in determining the router-level topology of ASs, and understanding and modeling the economic processes that determine bilateral peering relationships between ASs are important future research problems directed toward achieving this goal.

ACKNOWLEDGMENTS

This work was done at the University of Southern California's Information Sciences Institute, 4676 Admiralty Way, Suite 1001, Marina del Rey, CA 90292. This work was partially supported by the Defense Advanced Research Projects Agency under grant DABT63-98-1-0007. Any opinions, findings, and conclusions or recommendations expressed in this material are those of the authors and do not necessarily reflect the views of the Defense Advanced Research Projects Agency.

REFERENCES

[1] Burch, H., and B. Cheswick. "Mapping the Internet." *IEEE Comp.* **32(4)** (1999): 97–98.

[2] Cheswick, B., H. Burch, and S. Branigan. "Mapping and Visualizing the Internet." Paper presented at the USENIX 2000, held in San Diego, California, June, 2000.

[3] Claffy, K. C., and D. McRobb. "Measurement and Visualization of Internet Connectivity and Performance." June 2001. ⟨http://www.caida.org/tools/measurement/skitter/⟩.

[4] Cheng, C., R. Riley, S. Kumar, and J. Garcia-Lunes-Aceves. 'A Loop-Free Extended Bellman-Ford Routing Protocol without Bouncing Effect." In ACM SIGCOMM Computer Communications Review, vol. 19(4), 224–236. New York: ACM Press, 1989.

[5] Faloutsos, C., M. Faloutsos, and P. Faloutsos. "What does Internet look like? Empirical Laws of the Internet Topology." In *Proceedings of ACM SIGCOMM 1999*, Boston, MA. New York: ACM Press, 1999.

[6] Gao, L. "Inferring Autonomous System Relationships in the Internet." In *IEEE/ACM Transactions on Networking*, vol. 2(6), 733–745. New York: ACM Press, 2001.

[7] Govindan, R., and H. Tangmunarunkit. "Heuristics for Internet Map Discovery." In *Proceedings of IEEE INFOCOM 2000*, 1371–1380. Washington, DC: IEEE Press, 2000.

[8] Khanna, A., and J. Zinky. "A Revised ARPANET Routing Metric." In *ACM SIGCOMM Computer Communications Review*, vol. 12(4), 45–56. New York: ACM Press, 1989.

[9] Kleinrock, L., and F. Kamoun. "Hierarchical Routing for Large Networks: Performance Evaluation and Optimization." *Comp. Net.* **1** (1977): 155–174.

[10] McQuillan, J. M., I. Richer, and E. C. Rosen. "The New Routing Algorithm for the ARPANET." *IEEE Trans. Comm.* **7(15)** (1980): 1–7.

[11] Paxson, V. "End-to-End Routing Behavior in the Internet." In *Proceedings of the ACM SIGCOMM Conference on Applications, Technologies, Architectures, and Protocols for Computer Communications*, 25–38. New York: ACM Press, 1996.

[12] Paxson, V. "End-to-End Internet Packet Dynamics." In *Proceedings of the ACM SIGCOMM '97 Conference on Applications, Technologies, Architectures, and Protocols for Computer Communication*, vol. 27(4), 139–154. New York: ACM Press, 199.

[13] Phillips, G., H. Tangmunarunkit, and S. Shenker. "Scaling of Multicast Trees: Comments on the Chuang-Sirbu Scaling law." In *Proceedings of the conference on Applications, Technologies, Architectures, and Protocols for Computer Communication (SIGCOMM '99)*, 141–151. New York: ACM Press, 1999.

[14] Savage, S., A. Collins, E. Hoffman, J. Snell, and T. Anderson. "The End-to-End Effects of Internet Path Selection." In *Proceedings of the Conference on Applications, Technologies, Architectures, and Protocols for Computer Communication (SIGCOMM '99)*, 289–299. New York: ACM Press, 1999.

[15] Tangmunarunkit, H., R. Govindan, and S. Shenker. "Internet Topology: Discover and Policy Impace." This volume.

Hidden Failures: The Role of the Protection System in Major Disturbances in Power Systems

James S. Thorp
Jie Chen

1 INTRODUCTION

The bulk power system is of current popular interest because of deregulation and some of the resulting regional problems. It is also important to note that inexpensive and reliable power has been a foundation of economic growth for decades. Indeed, the National Academy of Engineering found that the electrification of the country was the greatest engineering accomplishment of the twentieth century. The power industry is large, with annual revenues of $220 billion, and keeps growing. It was the largest man-made system prior to the Internet. There are three large, tightly interconnected systems in the U.S.: the eastern interconnection, the western interconnection, and Texas. The east and west are connected at the Rocky Mountains, but the connection is relatively weak.

The power system is made up of generators, transmission lines, and a distribution network. The transmission system has voltages from 800 kV to 115 kV and the distribution system has voltages below 115 kV. The distribution system extends to virtually every home and business in the country. The structure and

The Internet as a Large-Scale Complex System,
edited by Kihong Park and Walter Willinger, Oxford University Press.

organization of transmission systems versus distribution systems are similar to those of the public switched telephone network (PSTN) versus local exchanges in the telephone systems. Generators can be as large as 1300 MW. There are tens of thousands of nodes (or buses) in the network in which transmission lines, transformers, and generators form the branches that are interconnected at the nodes. A dynamic system model used for stability studies could involve hundreds of thousands of ordinary differential equations describing the generators, their control systems, the loads, and the transmission system.

Except for a small amount of pumped hydro, there is essentially no storage in the power system. Daily, weekly, and seasonal load demands are predicted by various schemes (temperature is an important input) in order to have sufficient generation in place. All of the generators in the major interconnections (the eastern interconnection, the western interconnection, and Texas) are synchronized and run at the same frequency (60 Hz). Mismatches between the generation and the load (the power delivered to the customers) and losses (mainly resistance-heat-loss in lines and generators) cause the synchronized system to slow down if generation is inadequate and speed up if generation is too large. There is roughly a constant of proportionality between power and frequency. For example, in the eastern interconnection, an error of approximately 6000 MW corresponds to a frequency difference of 0.1 Hz. And typically, the allowed frequency deviation is $\pm 0.1 \sim 0.5$ Hz for systems of different sizes. To avoid damage to equipment caused by under-frequency operation, the load is shed if the frequency is too low. This process usually occurs in steps based on the measured frequency. Load shed by this "under-frequency load-shedding" is also restored based on the measured frequency when the frequency returns to near normal. A single global quantity, frequency, is thus used in a local control scheme to regulate the balance of power in the system.

2   THE PROTECTION SYSTEM

The bulk power system must be protected from short circuits in the same way that household wiring is protected from short circuits. Fuses, circuit breakers, and ground fault detectors are used in both. Fuses are typically limited to distribution systems, while the operation of circuit breakers is more complicated than in the home. Relays control the opening and subsequent reclosing of circuit breakers. Relays have inputs from local current and voltage transformers and from communication channels connected to the remote ends of the lines. The relay output trip signal is sent to circuit breakers. As an example, the most common source of "faults" or short circuits in the transmission system is lightning. A lightning strike near a high voltage line produces an ionized path to ground that supports a large, damaging current to flow. Fault currents can be as large as tens of thousands of amps. The circuit breaker is only able to interrupt this large current by opening a mechanical connection as the alternating current goes through zero. The ionized arc dissipates when the line is deenergized and the breakers can reclose and return the line to service. The time interval for such "high speed" reclosing is typically 20 cycles or one third of a second. Because the transmission system is a three-phase system, the relays must deal with a number of possible faults. With three phases there are three phase conductors and a ground wire on the towers in addition to the actual ground below. A short circuit between any two of these caused by lightning, a tree, a failed insulator, etc., is a fault. Most faults are between a single conductor and ground (a phase to ground fault), but a protection system must be able to contend with all possible fault combinations. In addition, more than one three-phase line can be supported by one set of towers. In most situations, the relays must decide whether a fault is on the line the relay is protecting or whether the fault is on an adjacent line. In the former case, the circuit breaker should open, but in the latter case, the relay should not send a signal to the breaker. Signals from the remote end of the line can make this determination easier.

When the power system was created, the first relays were electromechanical devices with moving mechanisms that made contact. Solid-state devices with discrete electronic components were introduced and were ultimately followed by microprocessor-based relays. Most existing microprocessor-based relays are stand-alone devices. But, as other devices (recording and control devices) in the substation become digital, an integrated substation architecture is emerging. The Utility Communication Architecture (UCA) envisions a Local Area Network (LAN) in the substation for the various intelligent electronic devices and a router connecting the substation LAN to a fiber network constructed on the transmission line rights-of-way. This utility Intranet would have to deal with relaying communications with a higher priority than revenue or power quality data in order to provide the required relaying speed. It has been estimated that there are tens of thousands of substations and up to 500,000 relays in the North American interconnection.

3 PROTECTION PHILOSOPHY

The existing protection system, which has evolved for a vertically integrated utility, may not be the best now or in the future. Existing protection systems were designed to protect equipment and not the system. The power system was robust enough so that removing a transmission line or transformer from service caused no interruption in service to customers. Present-day relaying systems are also designed to be dependable at the cost of security. It should be recognized that a relay has two failure modes. It can trip when it should not trip (a false trip) or it can fail to trip when it should trip. The two types of reliability have been designated as "security" and "dependability" by protection engineers. Dependability is defined as the measure of the certainty that the relays will operate correctly for all faults for which they are designed to operate, while security is the measure of the certainty that the relays will not operate incorrectly. The existing protection system with its multiple zones of protection and redundant systems is biased toward dependability, i.e., a fault is always cleared by some relays [20, 28]. There are typically two primary protection systems often relying on different principles (one might depend on communications while the other uses only local information) and multiple backup systems that trip (with some time delay) if both primary systems fail to trip. The result is a system that virtually always clears the fault but as a consequence permits large numbers of false trips. When the system is highly stressed or operated in new modes brought about by re-regulation, the consequences of false trips can be serious.

4   BLACKOUTS AND HIDDEN FAILURES

The North American Electric Reliability Council (NERC) prepares an annual report of major disruptions in North America reporting approximately ten or twenty major events a year [18]. The U.S. Department of Energy (DOE) established requirements for reporting electric system emergencies. In order to be reported the disturbance must involve a certain number of customers or megawatts and be of a specified duration:

- For utilities with previous year's peak load greater than 3000 MW, the loss of 300 MW of load for more than 15 minutes.
- For utilities with previous year's peak load less than 3000 MW, the loss of 200 MW or 50% of load for more than 15 minutes.
- Load shedding more than 100 MW.
- Continuous interruption greater than 3 hours to 50,000 customers or more than 50% of total customers served.
- Voltage reduction greater than 3%.

Over a long interval, more than 70% of the major disturbances involved relaying systems: not necessarily as the initiating event but contributing to the cascading nature of the event [18, 26]. The 1965 Northeast Blackout was initiated by a relay tripping on load current, a number of relays failed to trip in the 1977 New York City (NYC) Blackout, and there were incorrect relay operations in the multiple disturbances in the western U.S. in the summer of 1996.

One manifestation of the backup systems described above is to divide the area the relay protects into zones. The first zone is the amount of the line for which the relay provides fast-primary protection. The second zone is the part of the system in which the relay provides backup protection and the third zone protection amounts to backup of the backup. In addition to third zone trips there were incorrect generator trips resulting in the loss of needed generation. Low voltages led to imbalance in exciter SCR bridge circuits that incorrectly tripped generators.

The causes of the incorrect relay operations in the NERC reports were:

- application (29%) (the wrong relay or relaying philosophy used);
- maintenance (42%);
- setting (10%);
- calibration; and
- transient and others.

An example of the maintenance issue is the fact that one of the relays involved in the 1977 NYC blackout had been maintained a few weeks before the event. The last act in the maintenance involved pushing a contact that was bent in the test and inadvertently left damaged. Many of the incorrect relay operations can be characterized as "hidden failures" in that, since relays are only called upon in unusual circumstances, there can be something wrong in the relay that is hidden until there are faults or heavy loads near the relay. Defects that are so serious that the relay would misoperate immediately when it was returned to service after maintenance are not "hidden." Unfortunately, hidden failures seem to account for a number of relay operations involved in major cascading outages. It has been observed that other large systems also have protection systems. Financial systems actually have "circuit breakers." The human immune system is a very sophisticated protection system, and most of the symptoms of a cold are the results of this protection system. The congestion detection and handling mechanism in the TCP protocol could also be thought of as a "traffic breaker" in the Internet.

5 IMPACT OF PROTECTION SYSTEMS

In spite of its importance, the impact of protection system malfunctions on overall system reliability has not been well studied. A part of the problem is that major disturbances are rare. While individual blackouts have been studied in great detail, there have been no simulation tools to simulate large numbers of disturbances. For example, there has been no qualitative evaluation of the effect on system reliability of digital relay self-checking and monitoring. The NERC database for the last 15 years has only about 300 events. In addition to studying the few actual blackouts, we would like to examine many cascading outages leading to blackouts and create a larger database of simulated disturbances. To examine the simulation process, consider a typical hidden failure mechanism associated with a particular relaying scheme. The direction comparison carrier-blocking scheme is shown in figure 1.

The relays are directional in that any fault (including those beyond the line) in the direction shown will be seen by the relay. The communication channel is used to send a blocking signal so that if relay one, for example, sees a fault and does not receive the blocking signal it opens the breaker. Historically the

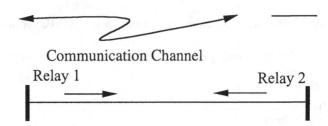

FIGURE 1   Direction comparison carrier-blocking.

communication channel was the power line itself and the blocking signal was a pure carrier in the 100 kHz range.

Hidden failure modes for this scheme are communication system failures that result in the loss of the blocking signal. Consider the portion of a power system shown in figure 2(a). We assume that the faulted line in figure 2(a) will be cleared by correct relay operation. Each of the arrows in the figure represents a directional comparison relay that "sees" the fault and must receive a blocking signal in order to block tripping. These relays are exposed to hidden failure trips because they are connected to the buses at the ends of the faulted line. With some small probability, p, a hidden failure in one of the communication systems could lead to a hidden failure trip of one of these lines. Such a possibility is shown in figure 2(b). Again, the arrows represent directional comparison relays which "see" the fault and must receive a blocking signal in order to block tripping. Again, with probability p, one of the exposed lines (in fig. 2(b)) could trip and might overload another line causing an overload trip, producing the system in figure 2(c).

It must be recognized that the flows in the remaining lines are governed by Kirchhoff's laws, i.e., there is no local control of the current (hence power) flowing in the lines. Once a line is removed by the circuit breakers, new flows will take over. There are many types of hidden failures depending on the relaying scheme (there is a large number of schemes). A disturbance can be described by a sample path made up of a sequence of hidden failure trips and correct trips due to overloads with resulting load and generation shedding.

The probability of a hidden failure, p, is taken as being independent for the sole purpose of the illustration above. More generally, each line will have a different flow-dependent probability of tripping incorrectly [1, 2, 30, 31]. The line is more likely to trip because of a hidden failure if the line is heavily loaded. The probabilities are small enough that multiple hidden failure trips at one branch almost never happen. The sample path is approximately one-dimensional, spreading in the system like a crack rather than two-dimensionally, like a forest fire.

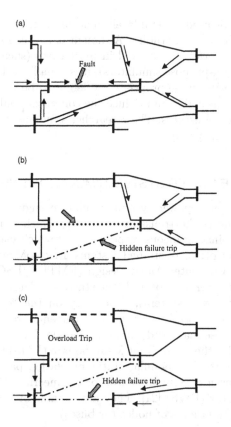

FIGURE 2   An illustration of a typical power system disturbance. (a) Fault on a line (the initiating event). (b) Fault-clearing line trip and the 1st hidden failure trip (cumulative path probability p). (c) Second hidden failure trip (cumulative path probability p^2) and another trip due to overload.

6   NUMERICAL SIMULATION OF POWER SYSTEM BLACKOUTS

The electric power system is a very complicated system consisting of tremendous sub-systems of great complexity themselves, such as computers, protection relays, and other electronic controlling appliances. Generally, it is necessary to model all the components playing an active role in the studied phenomena. Nevertheless, this statement encompasses several major difficulties: Which component should be modeled in detail? What parts of the network can be replaced by equivalents? Which component needs the dynamic model and which one requires only the static model? Furthermore, even though the decision has been made on which model should be used, several levels of accuracy may be envisaged depending on the aims of individual studies. Although it is a common practice to use models with enough details to be valid in most cases (such as commonly used models in most general-purpose simulators), it is usually still the expertise of the researchers that decides the appropriate modeling to be used, in order to keep the modeling effort reasonable and reduce the computational requirements, thus increasing speed and robustness. The researcher's judgment is always needed at the various steps of this process.

6.1   LARGE-SCALE POWER SYSTEM SIMULATION ENGINES

Common general-purpose power system simulation engines can be categorized into two groups. One is the transient simulator, which is mainly for simulating the time domain instantaneous responses, and which usually needs detailed electromagnetic descriptions of studied systems. ElectroMagnetic Transient and DC/Power Systems Computer Aided Design (EMTDC/PSCAD) [16] is an excellent transient simulator. The other is the static simulator aimed at simulating the steady-states of power systems, which rely on the "quasi-sinusoidal" approximation according to which voltages and currents are assumed to be pure sinusoidal signals, whose magnitude and phase angle change slowly with respect to 60 Hz. This justifies the use of Root Mean Square (RMS) values, phasors, real and reactive power, etc. The popular static simulation packages include (non-exhaustively), the Power System Simulator for Engineering (PSS/E) [24], and the Power World Simulator (PWS) [21]. They are capable of simulating large-scale power systems with thousands of nodes (or buses).

6.2   OPTIMAL POWER FLOW (OPF) SOLVERS

For the specific studies of hidden failures [1, 2, 27, 30, 31], the steady-state simulation is used. The core calculation required is to solve load flows (or power flows) subject to an objective function and network constraints (such as transmission line thermal limits, generator limits, voltage magnitude, and angle limits). This is often referred to as OPF. The objective function of OPF can take different forms, for example, minimization of the fuel cost, minimum shift of generation and other controls from an optimum operating point, or minimum load-shedding under disturbances. Most modern OPF solvers use the Newton-Raphson method [33] with the techniques of sparsity programming and optimally ordered Gaussian elimination [19, 25, 29], which have the advantage of strong convergence, low computer storage, and high computational speed. Interior-Point-Method-based nonlinear programming [33] is another well developed method for OPF, but it created only academic interest and has not been widely used by industrial users. Linear Programming OPF (LPOPF) [33] is another fully developed and commonly used method with the nonlinear objective functions and constraints handled by linearization. LPOPF can provide fast and robust solutions, and is especially capable of dealing with very large scale systems.

6.3   RARE-EVENT SIMULATION TECHNIQUES

6.3.1   Importance Sampling.

NERC reports [18] shows that there are only about 300 events during the past 16 years in the North American power system. A direct simulation of these rare events would require an unrealistically huge amount of computation [4]. One way out of this quandary is to use the importance sampling technique. In importance sampling, basically, rather than using the actual probabilities, the simulation uses altered probabilities (usually much bigger than actual ones) so that the rare events occur more frequently [4, 27]. Associated with each distinct sample path, SP_i, a ratio of actual probability of the event p_i^{actual} divided by the altered probability $p_i^{simulate}$ is computed. The estimated probability of SP_i then can be formed as

$$\hat{\rho}_i = \frac{N_{\text{occurring}}}{N_{\text{total}}} \times \frac{p_i^{actual}}{p_i^{simulate}} \qquad (1)$$

where $N_{\text{occurring}}$ is the number of times that SP_i occurred and N_{total} is the total number of samples. The mean value of $\hat{\rho}_i$ is unbiased. Even N_{total} is much smaller than the number otherwise required for direct Monte Carlo simulations [4]. A variation of the basic importance sampling technique was applied to achieve even better computational efficiency [27].

6.3.2   Heuristic Random Search.

Although importance sampling can significantly speed up the rare-event simulations, it still spends most of the computation resources in generating repeated samples to maintain the unbiased probability estimation. Another algorithm, referred to as the Heuristic Random Search [30, 31], was devised to search the important sample paths more efficiently. Each disturbance corresponds to a sample path made up of hidden failure trips and trips due to overloads. This algorithm concentrated on tracking only important sample paths and reducing the repetitions in the simulated samples as much as possible. One measurement of the importance of a sample path is the product of blackout size and probability. Bigger numbers indicate more significant disturbances. The algorithm used a mixing of random walk and greedy search based on the observation that power system disturbances spread in a one-dimensional fashion, i.e., more than one hidden failure is seldom triggered at the same time. The algorithm was a Depth First Searchs (DFS)-like search. Basically, the search is Depth-First, but at each branch point the sub-path with higher probability will be explored first. And the search will stop if the abovementioned measurement is too small (the actual number is system-dependent). Once it stops, no matter whether or not it reaches the leaf of the "blackout" tree, the search will restart from the root and try to find another "unvisited" important sample path. The procedure is repeated until the rightmost both of the "blackout" tree is reached (suppose the search starts from the leftmost).

6.4   PARALLEL COMPUTING

With the advance of computing technology, further simulation-speedup can be achieved by harnessing distributed computing. Efforts have been made in the literature [15] to develop parallel OPF algorithms and promising results have been obtained. But so far, widely used OPF approaches all face the hurdles of decomposition and factorization in parallel applications. It is very costly to adapt current simulation packages for parallel computing from both software and hardware viewpoints. However, a simple and low cost parallel solution does exist for non-sequential (i.e., no time correlations) simulations. Basically, the non-sequential simulation can be divided into a bunch of independent tasks with each of them carried out on different individual slave processors and one master processor responsible for communication and result collecting. The work of Wang and Thorp [30, 31] is an example of employing this idea.

6.5   MODIFIED HIDDEN FAILURE MECHANISM

A hidden failure mechanism is essential to the simulation of power system blackouts. The previous hidden failure model [1, 2] can be improved with a small modification. Suppose a line is exposed multiple times during a cascading event. One would expect that, if relay misoperation occurs, it would be more likely to occur on the first exposure than in subsequent exposures. However, the previous version of the model allowed relay misoperation with equal probability on all the line exposures. The improved model reduces or simply zeros the probability of misoperation after the first exposure.

6.6   SIMULATION PROCEDURE

The simulation procedure of power system cascading disturbances is summarized here, details can be found in Bae and Thorp [1, 2] and Wang and Thorp [30, 31]. Briefly, the simulation begins by randomly choosing an initial line trip. This action exposes all lines connected to the ends of the initial line and also may overload lines. If one line flow exceeds its preset limit then the line is tripped. Otherwise, the hidden failure mechanism is applied to let the chosen exposed line trip. After each line trip, the line flow is recalculated and checked for violations in line limits. The process is repeated until the cascading event stops. As a final step, an optimal distribution of generation and load is calculated. The above simulation is repeated over an ensemble of randomly selected transmission lines as the initiating fault locations.

7 CASE STUDIES OF POWER SYSTEM BLACKOUTS

7.1 OPTIMAL LOCATIONS FOR PROTECTION SYSTEM ENHANCEMENT

After the summer of 1996, NERC proposed extensive improvements in all protection systems. The cost was estimated to be in the billions. Recognizing that the industry has limited resources for protection system upgrades, it is more appropriate to ask where to make limited investments in protection improvements. The premise is that by using the self-checking and self-monitoring capability of digital relays, the probability of hidden failures can be reduced. The goal is to determine locations in the system where protection is sensitive and develop tools that can assist in planning system upgrades.

A study was made [30, 31] of a New York Power Pool (NYPP) 3000-bus system including a sequence of full AC load flows simulating cascading disturbances based on a model of hidden failures in malfunctioning relays. The system included load shedding, transmission line limits, generator's VAR limit, remote controlled buses, phase shift transformers, and switched shunt elements. Heuristic Random Search and distributed computing were used to achieve computational efficiency. The simulation was performed on the Cornell NT Cluster and employed 60 processors for 10 hours. More than 100,000 sample paths were generated and reduced to about 10,000 with the largest expected power loss. Results showed that hidden failures in relays at some locations were more prone to triggering cascading disturbances than elsewhere. The number of times a given line appeared in the 10,000 sample paths or the expected energy loss in the sample paths with that line being involved can be used as performance indices. It was possible to identify locations where limited resources should be expended to reduce the probability of hidden failures. Since actual hidden failure probabilities are not known the improvement was taken to be a 50% reduction in hidden failure probability. The optimization found the ten locations where a reduction in hidden failure probability by 50% would have the greatest impact.

7.2 CRITICALITY VERSUS POWER SYSTEM LOADING LEVEL

Much faster simulations result if the hidden failure model is used with a DC load flow approximation, in which the linearized, lossless power system is equivalent to a resistive circuit with current sources. In particular, transmission lines may be regarded as resistors, and generation and load may be regarded as current sources and sinks at the nodes of the network. This approximation captures the general features of cascading outages in power systems and can be used to obtain fast results when studying complex system behaviors such as criticality. As one can imagine, in any power system, at zero loading there are no blackouts and at some absurdly large loading there is always a blackout. The nature of the transition between these two extremes is of great interest to researchers [10].

A 179-bus Western Systems Coordinating Council (WSCC) equivalent system was simulated by applying DC load flow approximation, importance sam-

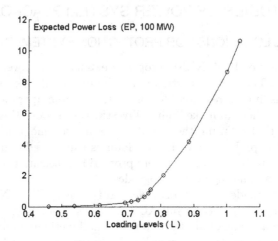

FIGURE 3   Variation of expected blackout size EP with loading.

pling technique, and LPOPF power dispatch [10]. The expected blackout size and blackout size Probability Distribution Function (PDFs) were examined as the load was increased.

Figure 3 shows the expected power loss

$$EP = \sum P_i \hat{\rho}_i \tag{2}$$

as a function of loading level L, where $\hat{\rho}_i$ is defined as in eq. (1), and P_i is the power loss associated with the ith sample path. The change in slope occurs near loading $L = 0.75$, which is 75% of the full system loading capacity. Figure 3 implies that as the system load increases a transition occurs and the system becomes more likely to have large disturbances. It is then plausible that the consequent risk of large blackouts due to hidden failures could be substantially reduced by lowering power system loading.

To find out the PDF of blackout size P, binning of the data was used. Assume that there are K sample points in bin j, and that each of the K points has the associated data pair $(P_i, \hat{\rho}_i)$. The representative $(\overline{P_i}, \overline{\hat{\rho}_i})$ for bin j is defined as

$$\overline{P_j} = \frac{1}{K} \sum_{i=1}^{K} P_i \tag{3}$$

$$\overline{\hat{\rho}_j} = \frac{\sum_{i=1}^{K} P_i \hat{\rho}_i}{\overline{P_j}} . \tag{4}$$

The variable binning is used here in such a way that each bin starts with the minimum scale and ends with at least a minimum number of samples. The

PDF then can be approximated by the binning data from eqs. (3)–(4). The similar binning techniques were also used in other studies to explore the power-law scaling behavior observed in major power system disturbances [6, 10], and produced various forms of power-law curves for relative frequency versus the size of the disturbance.

8 OTHER WORK

Recently researchers have examined the data on major disturbances in power systems [18]. In the past, investigators tended to focus on individual causes of these disturbances and assumed that the probability of the occurrence of blackouts decayed exponentially with the event size. This is in contrast to the recently found power-law "tail" of blackouts [6]. Many distributions of observed quantities in a wide variety of complex systems, such as earthquakes, forest fires, and even biological evolution, exhibit power-law form in their tails, sometimes called "heavy tailed distribution." During the last decades, books, conference proceedings, and papers have appeared concerning such power-law scaling behavior. Several different power-law-producing mechanisms have been proposed for power systems, among which the theory of self-organized criticality (SOC) [3, 14] is the most accepted and investigated. Recently, another idea, highly optimized tolerance (HOT) [5], was added to this toolkit, helping to explain power-law behavior in designed systems. Detailed models of power-transmission-networks with hidden failure models for the protection system have also produced similar power-law distributions [7, 10]. The influence model, a stochastic network model composed of interacting Markov chains, has also been used to study resource allocation in networks [23].

Project Eligible Receiver demonstrated that the computer systems that control the electric power grids are readily accessible to hackers, and that the grid is increasingly dependent on these systems for its health and efficiency [12, 17]. It must be recognized that the major blackouts that have been discussed in the preceding were accidental and a result of a combination of unlikely events such as hidden failures, but not malicious intervention. While Project Eligible Receiver focused on the control of the generation in the system, it is also possible to imagine hackers gaining access to the microprocessor-based relays and tripping significant transmission lines. The challenge is to design a modern protection system that is secure in the computer network sense.

9 FUTURE CHALLENGES

Normally, a protective system responds to faults or abnormal events in a fixed, predetermined manner. This predetermined manner, embodied in the characteristics of the relays, is based upon certain assumptions made about the power system. A new concept of "Adaptive Relaying" accepts that relays, which protect a power network, may need to change their characteristics to suit the prevailing power system conditions. With the advent of digital relays, the concept of responding to system changes took on a new dimension. Digital relays have two important characteristics that make them vital to the adaptive relaying concept. Their functions are determined through software and they have a communication capability, which can be used to alter the software in response to higher-level supervisory software or under the commands from a remote control center.

The ability to change a relay characteristic or setting, on the fly, as it were, raised serious questions about reliability and responsibility. Adaptive relaying with digital relays was introduced on a major scale in 1987 [22, 13]. One of the driving forces that led to the introduction of adaptive relaying was the change in the power industry wherein the margins of operation were being reduced due to environmental and economic restraints and the emphasis on operation for economic advantage. Consequently, the philosophies governing traditional protection and control performance and design are being challenged.

Various backup protection systems based on either expert systems or adaptive architectures have been proposed in an attempt to correct for the traditional relay system's shortcomings [8, 9, 32]. A promising research area is the use of agent-based backup protection systems. In this context, an agent is a self-contained piece of software that has the properties of autonomy and interaction. A backup protection proposal using geographically distributed agents located in every relay can improve on the more traditional isolated component system. In order to make this distributed system work effectively, the relays must be capable of autonomously interacting with each other. An agent makes decisions without the direct intervention of outside entities. This flexibility and autonomy adds reliability to the protection system because any given agent-based relay can continue to work properly despite failures in other parts of the protection system. It is certainly necessary to explore the expected communication traffic patterns in order to make agents more intelligent and robust towards network conditions. By linking the most widely used computer network simulator NS2 to an accurate power systems simulation engine EMTDC/PSCAD, the group developed and evaluated a family of new power grid protection and monitoring algorithms. Simulation reveals a surprising issue: in TCP-based power systems communication networks, relay protection algorithms may malfunction if TCP's congestion-control mechanisms are triggered (e.g., because of competing network traffic) [8, 9, 32].

A major contributor to cascading outages is the limited range over which information is communicated to define the real-time state of the system topology and parameters. Measurements of voltages and currents at various buses in a power system define the state of the power system. If several quantities are to be measured and transmitted to a central location to analyze the real-time state of the system, it is essential that they be measured with a common reference. The reference can be determined by the instant at which the samples are taken. In order to achieve a common reference for the measurements, it is essential to achieve synchronization of the sampling pulses. A time stamp, provided by a Global Positioning System (GPS) receiver clock, constitutes a part of the data communicated about each measurement. Several experimental systems utilizing these ideas are being installed in some power systems. The Electric Power Research Institute (EPRI) roadmap [11] envisions the rapid deployment of a Wide-Area Measurement System (WAMS) based on the GPS technology. Such a system is in the early stages of development in the WSCC. Various versions of the measurement units are interconnected by data concentrators and communicated to a few central locations. Such systems are planned for the entire western system.

10   CONCLUSION

The role of the protection system in major disturbances in large electric power systems has been investigated and simulation tools for studying such rare events have been presented. While not all large-interconnected systems have the same kind of protection system, some of the concepts developed here seem to have general relevance. The idea of a "hidden failure" in the system, i.e., a flaw in design or application or an error in maintenance that does not cause an immediate response but which can, when the system is stressed, exacerbate conditions to produce a cascading event, seems to apply to many systems including the Internet. Hidden failures are associated with systems that are inherently robust in that the system continues to function acceptably with a number of hidden failures until a large disturbance exposes hidden failures and becomes a much larger disturbance. In some sense, the hidden failures can be thought of as part of the mechanism for producing the heavy tails in the power-law curves.

REFERENCES

[1] Bae, K., and J. S. Thorp. "An Importance Sampling Application: 179 Bus WSCC System under Voltage Based Hidden Failures and Relay Misoperations." *Proceedings of the 31st Hawaii International Conference on System Sciences*, edited by R. H. Sprague, Jr., vol. 3, 39–46. Computer Society Press, 1998.

[2] Bae, K., and J. S. Thorp. "A Stochastic Study of Hidden Failures in Power System Protection." *Decision Support Systems* **24** (1999): 259–268.

[3] Bak, P., C. Tang, and K. Wiesenfeld. "Self-Organized Criticality: An Explanation of $1/f$ Noise." *Phys. Rev. Lett.* **59** (1987): 381–384.

[4] Bucklew, J. A. *Large Deviation Techniques in Decision, Simulation, and Estimation*. John Wiley and Sons, Inc. 1990.

[5] Carlson, J. M., and J. Doyle. "Highly Optimized Tolerance: A Mechanism for Power Laws in Designed Systems." *Phys. Rev. E* **60** (1999): 1412–1427.

[6] Carreras, B. A., D. E. Newman, I. Dobson, and A. B. Poole. "Initial Evidence for Self-Organized Criticality in Electric Power System Blackouts." *Proceedings of the 33rd Hawaii International Conference on System Sciences* (CD-ROM), January 4-7, 2000. Computer Society Press, 2000.

[7] Chen, J., J. S. Thorp, and M. Parashar. "Analysis of Electric Power Disturbance Data." *Proceedings of the 34th Hawaii International Conference on System Sciences* (CD-ROM), January, 2001. Computer Society Press, 2001.

[8] Coury, D. V., J. S. Thorp, K. M. Hopkinson, and K. P. Birman. "An Agent-based Current Differential Relay for use with a Utility Intranet." *IEEE Trans. Power Del.* **17(1)** (2002): 47–53.

[9] Coury, D. V., J. S. Thorp, K. M. Hopkinson, and K. P. Birman. "Improving the Protection of EHV Teed Feeders Using Local Agents." *Proceedings of IEEE Seventh International Conference on Developments in Power System Protection*. Washington, DC: IEEE Press, 2001.

[10] Dobson, I., J. Chen, J. S. Thorp, B. A. Carreras, and D. E. Newman. "Examining Criticality of Blackouts in Power System Models with Cascading Events." *Proceedings of the 35th Hawaii International Conference on Systems Sciences* (CD-ROM), January 2002. Computer Society Press, 2002.

[11] EPRI Electricity Technology Roadmap: Power Delivery Module. Draft, October 15, 1998.

[12] Gertz, B. "'Infowar' Game Shutdown U.S. Power Grid, Disabled Pacific Command." *Washington Times* **Apr. 16** (1998): A1.

[13] Horowitz, S. H., A. G. Phadke, and J. S. Thorp. "Adaptive Transmission System Relaying." *IEEE Trans. Power Del.* **3(4)** (1988): 1436–1445.

[14] Jensen, H. J. *Self-Organized Criticality: Emergent Complex Behaviour in Physical and Biological Systems*. Cambridge, MA: Cambridge University Press, 1998.

[15] Kim, B. H., and R. Baldick. "Comparison of Distributed Optimal Power Flow Algorithms." *IEEE Trans. Power Sys.* **15(2)** (2000): 599–604.

[16] Manitoba HVDC Research Centre. "Getting Started." In *EMTDC/PSCAD Manual, version 3.0*, chapter 5, 95 pages. June 2001. ⟨http://ee.changwon.ac.kr/sinposylab/image/pscad/⟩.

[17] Myers, L. "Pentagon Has Computers Hacked." *Assoc. Press* **Apr. 16** (1998).

[18] North American Electric Reliability Council. "Disturbances Analysis Working Group Database." 20004. ⟨http://www.nerc.com/~dawg/database.html⟩v.

[19] Ogbuobiri, E. C., W. F. Tinney, and J. W. Walker. "Sparsity-Directed Decomposition for Gaussian Elimination on Matrices." *IEEE Trans. Power Apparatus and Sys., PAS-89* **1** (1970): 141–150.

[20] Phadke, A. G., and J. S. Thorp. "Expose Hidden Failures to Prevent Cascading Outages in Power Systems." *IEEE Comp. App. Power* **9** (1996): 20–23.

[21] Power World Corporation, the Visual Approach to Analyzing Power Systems. Home Page. October 2004. ⟨http:/www.powerworld.com/products/simulator.html⟩.

[22] Rockefeller, G. D., C. L. Wagner, J. R. Linders, K. L. Hicks, and D. T. Rizy. "Adaptive Transmission Relaying Concepts for Improved Performance." *IEEE Trans. Power Del.* **3(4)** (1988): 1446–1458.

[23] Roy, S., B. C. Lesieutre, and G. C. Verghese. "Resource Allocation in Networks: A Case Study of the Influence Model." *Proceedings of the 35th Hawaii International Conference on Systems Sciences* (CD-ROM), January 2002. Computer Society Press, 2002.

[24] Shaw Power Technology, Inc. PTI Home Page. October 2003. ⟨http://www.shawgrp.com/PTI/index.cfm⟩.

[25] Stott, B., and E. Hobson. "Solution of Large Power-System Networks by Ordered Elimination: A Comparison of Ordering Schemes." *Proc. IEEE* **118(1)** (1971): 125–134.

[26] Tamronglak, S., S. H. Horowitz, A. G. Phadke, and J. S. Thorp. "Anatomy of Power System Blackouts: Preventive Relaying Strategies." *IEEE Trans. Power Del.* **11** (1996): 708–715.

[27] Thorp, J. S., and A. G. Phadke. "Anatomy of Power System Disturbances: Importance Sampling." Proceedings of the 12th Power Systems Computational Conference (PSCC), Dresden, Germany, Aug. 1996.

[28] Thorp, J. S., A. G. Phadke, S. H. Horowitz, and S. Tamronglak. "Anatomy of Power System Disturbances: Importance Sampling." *Intl. J. Elec. Power & Energy Sys.* **20(2)** (1998): 147–151. Also presented at the 12th Power Systems Computation Conference, Dresden, Germany, August 19-23, 1996.

[29] Tinney, W. F., and J. W. Walker. "Direct Solutions of Sparse Network Equations by Optimally Ordered Triangular Factorization." *Proc. IEEE* **55(11)** (1967): 1801–1809.

[30] Wang, H., and J. S. Thorp. "Enhancing Reliability of Power Protection Systems Economically in the Post-Restructuring Era." Proceedings of the 32nd North American Power Symposium NAPS). July 2000. ⟨http://www.pserc.org/ecow/get/publicatio/2000public/⟩.

[31] Wang, H., and J. S. Thorp. "Optimal Locations for Protection System Enhancement: A Simulation of Cascading Outages." *IEEE Trans. Power Del.* **16** (2001): 528–534.

[32] Wang, X. R., K. M. Hopkinson, J. S. Thorp, R. Giovanini, K. P. Birman, and D. V. Coury. "Developing an Agent-based Backup Protection System for Transmission Networks." Submitted to the Conference on Power Systems and Communication Systems Infrastructures for the Future, Beijing, China, September 2002. ⟨http://www.cs.cornell.edu/hopkik/protection_agent.pdf⟩.

[33] Wood, A. J., and B. F. Wollenberg. *Power Generation Operation and Control.* 2d ed. John Wiley & Sons, Inc., 1996.

A Note on Statistical Multiplexing and Scheduling in Video Networks at High Data Rates

Jörg Liebeherr

This chapter makes observations about the impact of link scheduling and statistical multiplexing on the multiplexing gain at high data rates. The multiplexing gain is evaluated using the number of MPEG video traces that can be provisioned with delay guarantees on a network link. The presented data indicates that, at high transmission rates, the multiplexing gain is substantial and is dominated by the effects of statistical multiplexing.

1 INTRODUCTION

A distinguishing property of packet-switching networks is that they achieve multiplexing gain by sharing resources. We speak of multiplexing gain when providing a certain grade of service to a group of traffic flows requires fewer network resources per flow than providing the same grade of service individually to each flow.

The Internet as a Large-Scale Complex System,
edited by Kihong Park and Walter Willinger, Oxford University Press.   179

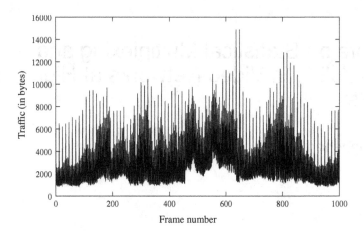

Traffic (in bytes) / Frame number

FIGURE 1   MPEG video traffic. Size of video frames for 1000 frames from the movie "Silence of the Lambs." The frame rate is 24 frames per second.

Research on quality-of-service (QoS) networks in the 1990s showed that multiplexing gain can be considerable even when service requirements of traffic are stringent and traffic is bursty, in the sense that the rate of traffic varies greatly over time. An example of a network service with stringent requirements is a bounded delay service, which guarantees that all traffic from a flow satisfies worst-case delay bounds and that no packets are dropped [20]. An example of bursty traffic is MPEG-1 video traffic, which has correlations of traffic over multiple time scales [21]. The burstiness of MPEG traffic is illustrated in figure 1 for a sequence from an MPEG-1 compressed motion picture. In Wrege et al. [60], it was shown that packet networks with a bounded delay service for MPEG-1 video traffic can capture multiplexing gain.

There are several approaches to improve multiplexing gain in a packet-switching network with service guarantees. For example, buffering traffic which exceeds a given burstiness constraint at the network entrance effectively smoothes the traffic rate [67]. However, due to possibly significant delays, smoothing is generally reserved for applications with a high delay tolerance. Scheduling algorithms at the output links of packet switches can improve multiplexing gain by transmitting backlogged traffic in the order which best satisfies the service requirements. One can also extract multiplexing gain by exploiting statistical properties of traffic. This is referred to as statistical multiplexing gain. By allowing a fraction of traffic to violate its service guarantees, for example, by making probabilistic service guarantees of the form: $Pr[Delay > X] < \varepsilon$, where ε is small, it is feasible to exclude worst-case traffic arrival scenarios where sharing of resources is difficult to achieve.

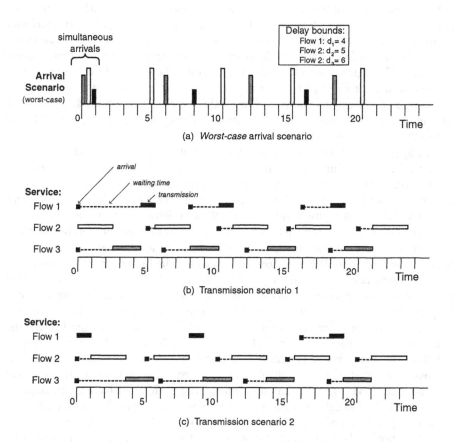

FIGURE 2 Multiplexing gain through scheduling. In figure 2(a), packet arrivals are indicated as boxes, where the color of a box indicates the flow type, and the length of the box indicates the transmission time of a packet. The transmission times for Flows 1, 2, and 3 are given by 1, 2.5, and 2 time units, respectively. The delay bounds for packets are given by $d_1 = 4$ for Flow 1, $d_2 = 5$ for Flow 2, and $d_3 = 6$ for Flow 3. We assume that packets from Flow 1, Flow 2, and Flow 3 arrive at most every eight, five, and six time units, respectively. In the depicted "worst-case" arrival scenario, packet arrivals from all flows coincide at time $t = 0$. In (b) and (c), two transmission scenarios are presented. Dotted lines present waiting times and the horizontal boxes indicate packet transmissions. In the transmission scenario in (b), packets are transmitted in the order of their arrival. This results in a violation of the delay bound of Flow 1 at time $t = 4$. The scenario in (c) gives higher priority to packets with shorter delay bounds. Here, no delay bound violation occurs.

We refer to figure 2 for a simple arrival scenario of three traffic flows. The arrival scenario in figure 2(a) depicts a *worst-case* scenario with simultaneous arrivals from all flows at time $t = 0$. In figures 2(b) and (c), two transmission scenarios are presented. In the first scenario in figure 2(b), packets are transmitted in the order of arrivals (and in some arbitrary order for simultaneous arrivals). The figure shows that this scheduling scheme results in a violation of a delay bound for Flow 1 at time $t = 4$. In figure 2(c), the transmission schedule gives highest priority to the packet with the smallest delay bound. This scheduler does not have delay bound violations. Thus, the scheduling algorithm in figure 2(c) yields a better multiplexing gain.

In figure 3 we show an example which intends to demonstrate the benefits of statistical multiplexing. Figure 3(a) shows a "typical" arrival scenario where simultaneous arrivals of packets are relatively rare. By excluding these rare events, a schedule that transmits packets in the order of arrivals, which resulted in delay bound violations in figure 3, does not result in a delay bound violation.

Previous work on video networks with QoS has focused on methods to characterize a video flow, for example, using leaky or token buckets [25, 36, 41, 46, 52, 53]. The impact of the selection of the packet scheduling algorithm was investigated in Krunz and Tripathi [37] and Wrege et al. [60], where it was shown that packet scheduling can have a noticeable impact on the number of flows that can be provisioned with a deterministic delay service. Several studies have explored the benefits of reducing the burstiness (*smoothing*) of video traffic by shaping traffic at the network entrance [23, 34] or by buffering traffic at the receiver [19, 67]. The above works generally consider worst-case traffic scenarios, and do not account for statistical multiplexing of video flows. Research on statistical multiplexing of video traffic has determined the buffer and bandwidth requirements needed to satisfy a certain quality of service, generally by ignoring the impact of packet scheduling [18, 21, 27, 28, 35, 51]. A direct comparison of the impact of both scheduling and statistical multiplexing has not been done, due to the lack of analytical tools that enable such a comparison. With a recently presented method to determine the statistical multiplexing gain with various scheduling algorithms [4], such a comparison has become feasible.

This chapter discusses a set of numerical examples from a recent technical report [3] which use the approach in Boorstyn et al. [4] to evaluate the number of video flows that can be provisioned with delay guarantees on a network link. The examples show that at high data rates, statistical multiplexing gain dominates the multiplexing gain, and, in comparison, the multiplexing gain due to scheduling is modest. Since the statistical multiplexing gain is due to the nature of traffic and is not part of the network design, the results indicate that a relatively simple QoS network design may achieve a high multiplexing gain.

The remaining sections of this chapter are structured as follows. In section 2 we specify our assumptions on the traffic and present schedulability conditions for a general class of packet schedulers. In section 3 we discuss numerical examples

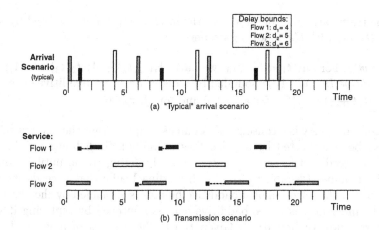

FIGURE 3 Statistical multiplexing. In a "typical" arrival scenario, packet arrivals do not coincide. A schedule which transmits packets in the order of arrivals does not result in a violation of delay bounds.

using MPEG video traces, and compare the multiplexing gain attainable through scheduling and through statistical multiplexing.

2 ANALYSIS FRAMEWORK

The evaluation of scheduling and statistical multiplexing in this chapter is based on analytical methods and not on experimental measurements. Specifically, we use schedulability conditions for a service with delay guarantees, which verify whether a network link can satisfy delay guarantees for a given set of traffic flows and a given scheduling algorithm. In this section, we discuss the schedulability conditions used in this chapter. We first present functions that describe arrivals of single and aggregate traffic flows, and then use these functions in the schedulability conditions.

2.1 TRAFFIC ARRIVALS AND ENVELOPE FUNCTIONS

We consider arrivals of groups of video flows to an output link of a packet switch with transmission rate C. At the link, a scheduler determines the order in which backlogged traffic is transmitted. Consider a set C of flows which are partitioned into Q flow types, where C_q denotes the subset of type-q flows. Delay guarantees for a video flow $j \in C_q$ are specified in terms of a delay bound d_q for type-j video flows. A delay bound violation occurs if traffic from flow j experiences a delay exceeding d_q.

The traffic arrivals from flow j in the interval (t_1, t_2) are denoted by a function $A_j(t_1, t_2)$ with the following properties:

- *Additivity*: For any $t_1 < t_2 < t_3$, we have $A_j(t_1, t_2) + A_j(t_2, t_3) = A_j(t_1, t_3)$.
- *Subadditive Bound*: A_j is bounded by a deterministic subadditive envelope A_j^* as $A_j(t, t + \tau) \le A_j^*(\tau)$ for all $t \ge 0$ and for all $\tau \ge 0$.[1]

The second property is a consequence of an assumption that the amount of traffic that can be transmitted by a video flow is limited. Such limits on the traffic can be enforced at the network entrance by buffering traffic until it conforms to a given bound ("shaping"), or by discarding traffic that exceeds its bound ("policing"). The selection of subadditive bounds is explained by the observation that a bound which is not subadditive can be improved by replacing it with a smaller subadditive bound [8]. Given the traffic arrivals of a video flow, the smallest subadditive bound for a traffic flow is given by

$$\mathcal{E}_j^*(\tau) = \sup_{t \ge 0} A_j(t, t + \tau) \qquad \forall \tau \ge 0 . \tag{1}$$

In Wrege et al. [60], this function is referred to as *empirical envelope*. To reduce the number of parameters of the empirical envelope, we can apply a method from Wrege et al. [60], which approximates the concave hull of the empirical envelope by a piecewise linear function with K segments. For a flow j, the kth segment of this function is characterized by a burst parameter σ_{jk} and a rate parameter ρ_{jk}, resulting in a subadditive envelope of the form

$$A_j^*(\tau) = \min_{k=1,\ldots,K} \{\sigma_{jk} + \rho_{jk}\tau\} , \tag{2}$$

where $\{\sigma_{jk}, \rho_{jk}\}_{k=1}^{K}$ are the parameters of the piecewise linear segments. An algorithm to reduce the number of piecewise linear segments can be found in Liebeherr and Wrege [41]. Envelopes with piecewise linear segments can easily be enforced by leaky buckets, a well-known group of traffic algorithms for policing or shaping of network traffic.

To exploit statistical multiplexing, we view the arrivals $A_j(t_1, t_2)$ as stochastic processes, defined over an underlying joint probability space that is suppressed in the notation. We require that the above assumptions hold for each sample path. In addition we assume that the following holds:

- *Stationarity*: The A_j are *stationary* so that for all $t, t' \ge 0$ we have $Pr[A_j(t, t + \tau) \le x] = Pr[A_j(t', t' + \tau) \le x]$.
- *Independence*: The A_i and A_j are stochastically independent for all $i \ne j$.

Within the constraints of these assumptions, we consider arrival scenarios where each video flow exhibits its worst possible ("adversarial") behavior. However,

---

[1] A function f is subadditive if $f(t_1 + t_2) \le f(t_1) + f(t_2)$, for all $t_1, t_2 \ge 0$.

even if flows individually behave in a worst-case fashion, as allowed by their subadditive bounds, the independence assumption prevents the flows from conspiring with each other to yield a joint worst-case behavior. These assumptions effectively exclude scenarios as shown in figure 2, where arrival bursts from multiple flows coincide.

For the calculation of statistical multiplexing gain we will take advantage of the notion of *effective envelopes* [4]. Effective envelopes are functions that are, with high probability, upper bounds on multiplexed traffic from a set of flows satisfying the given assumptions. Effective envelopes have been shown to be a useful tool for calculating the statistical multiplexing gain at a network node [6, 43].[2]

Consider the set $\mathcal{C}_q$ of type-q flows. We use $A_{\mathcal{C}_q}$ to denote the aggregate arrivals of all type-q flows, that is, $A_{\mathcal{C}_q}(t, t + \tau) = \sum_{j \in \mathcal{C}_q} A_j(t, t + \tau)$. Let N_q denote the number of flows in set $\mathcal{C}_q$. All flows of the same type have the same subadditive bound. Thus, we use A_q^* to denote the bound of a type-q flow with $A_j^*(\tau) = A_q^*(\tau)$ for all $j \in \mathcal{C}_q$.

An effective envelope for $A_{\mathcal{C}_q}(t, t + \tau)$ is a function $\mathcal{G}_{\mathcal{C}_q}$ with:

$$Pr\left[A_{\mathcal{C}_q}(t, t + \tau) \leq \mathcal{G}_{\mathcal{C}_q}(\tau; \varepsilon) \right] \geq 1 - \varepsilon, \quad \forall\, t,\ \tau \geq 0 . \tag{3}$$

Thus, an effective envelope provides a bound for the aggregate arrivals $A_{\mathcal{C}_q}$ for each time interval of length τ, which is violated with probability at most ε.

Explicit expressions for effective envelopes can be obtained with large deviation techniques. The construction of effective envelopes $\mathcal{G}_{\mathcal{C}_q}$ for a set $\mathcal{C}_q$ of type-q flows uses the moment generating function of A_j, denoted as $M_j(s, t) = E[e^{A_j(0,t)s}]$, where $E[.]$ denotes the expected value. As shown in Boorstyn et al. [4], with the above assumptions, $M_j(s, t) \leq \overline{M}_q(s, t)$ holds for each flow $j \in \mathcal{C}_q$, where

$$\overline{M}_q(s, t) = 1 + \frac{\rho_q t}{A_q^*(t)} \left(e^{s A_q^*(t)} - 1 \right) , \tag{4}$$

and where $\rho_q := \lim_{t \to \infty} A_q^*(t)/t$ is assumed to exist. With the independence of flows we obtain from the Chernoff bound that

$$Pr\{ A_{\mathcal{C}_q}(t) \geq x \} \leq e^{-xs} \overline{M}_q(s, t)^{N_q} . \tag{5}$$

From here, we can obtain an effective envelope as follows

$$\mathcal{G}_{\mathcal{C}_q}(t; \varepsilon) = \inf_{s>0} \frac{1}{s} \left(N_q \log \overline{M}_q(s, t) + \log \varepsilon^{-1} \right) . \tag{6}$$

Note that the effective envelope can also be expressed in terms of the *effective bandwidth* given by $\alpha(s, t) := 1/st \log M_j(s, t)$, which is widely used in the literature on statistical multiplexing [28].

---

[2]In Boorstyn et al. [4] two notions of effective envelopes are introduced, called local effective envelope and global effective envelope. In this chapter, we only use local effective envelopes and refer to them as effective envelopes.

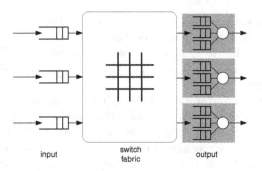

FIGURE 4   Packet switch.

We will use the effective envelope given by eq. (6) in the numerical examples in section 3. We will see that effective envelopes capture the statistical multiplexing gain well. If N_q is large, generally, we have that $\mathcal{G}_{\mathcal{C}_q}(t) \ll N_q \cdot A_q^*(t)$.

2.2 SCHEDULABILITY

In a packet-switched network, packets from a particular flow traverse the network on a path of packet switches and links. Figure 4 shows a sketch of a typical packet switch. The shown switch performs buffering at the input and the output of a switch fabric. For each output link, a scheduler determines the order in which backlogged packets are transmitted. The selection of a particular scheduling discipline for an output link presents a tradeoff between the need to support a large number of flows with diverse delay requirements and the need for simplicity of the scheduling operations. Here we consider well-known scheduling disciplines first-come-first-served (FCFS), static priority (SP), and earliest-deadline-first (EDF).

- *First-come-first-served (FCFS)* : A FCFS scheduler transmits packets in the order of their arrival. The main advantage of FCFS is its simplicity. However, since an FCFS scheduler treats all traffic in the same way, it is not well suited to support different delay guarantees.
- *Static priority (SP)* : An SP scheduler assigns to each flow type a priority level and a separate FCFS queue. Traffic is always transmitted from the highest priority FCFS queue. By convention, a lower index indicates a higher priority level.
- *Earliest-deadline-first (EDF)* : An EDF scheduler tags traffic with a deadline which is set to the arrival time plus the delay bound d_q, and transmits traffic in the order of deadlines. It has been shown that the EDF scheduling algorithm is optimal for a service with delay guarantees, in the sense that, among all

scheduling algorithms, it can support the most flows with deterministic delay guarantees [42].

Given a scheduling algorithm and a set of delay bounds, a *schedulability condition* verifies that, for all flows, the delay of each packet is less than its required delay bound. In the following, we will not take into consideration that packet transmissions on a link cannot be preempted. This assumption is reasonable when packet transmission times are short. We assume that the transmission rate of the link is normalized, that is $C = 1$.

In addition to the scheduling algorithm and statistical multiplexing, the number of flows that are admitted by a schedulability condition depends on the method in which traffic is characterized in the schedulability condition, and on the tightness of the schedulability condition itself. If a traffic characterization method overestimates the actual traffic of a flow, the schedulability condition will underestimate the achievable multiplexing gain. Since schedulability conditions are expressed in terms of bounds, loose bounds have the same effect. The traffic characterization used in this chapter, is based on the empirical envelope of an MPEG trace. This method has been demonstrated to be very accurate and cannot be easily improved [41]. Likewise, the schedulability conditions presented here, from Boorstyn et al. [4], are relatively tight.

A schedulability condition for a set of flows with deterministic delay guarantees is given as follows. If a set of flows satisfies the assumptions on additivity and subadditive bounds, no delay violation occurs if the d_q are selected such that, for all $\tau \geq 0$, we have

$$\sup_{\tau} \left\{ \sum_p \sum_{j \in C_p} A_j^*(x_p^{q,\tau} + \tau) - \tau \right\} \leq d_q , \qquad (7)$$

where $x_p^{q,\tau}$ is given by

$$\text{FCFS:} \quad x_p^{q,\tau} = 0$$

$$\text{SP:} \quad x_p^{q,\tau} = \begin{cases} -\tau & , \quad p > q \\ 0 & , \quad p = q \\ d_q & , \quad p < q \end{cases}$$

$$\text{EDF:} \quad x_p^{q,\tau} = \max\{-\tau, \ d_q - d_p\} .$$

To illustrate the conditions, suppose that time 0 is the last time that the scheduler was empty. Then, the condition implies that the delay bound d_q is large enough so that, for a tagged traffic arrival at time τ, the traffic that will be transmitted before the tagged arrival does not cause the arrival to exceed its delay bound d_q. For FCFS, the maximum amount of traffic that will be transmitted before the tagged arrival at time τ is given by $\sum_{j \in C_p} A_j^*(\tau) - \tau$, that is, the difference between the maximum traffic arrivals before the arrival of the tagged arrival,

$\sum_{j \in \mathcal{C}_p} A_j^*(\tau)$, and the amount of traffic that has been transmitted since time 0, τ. The conditions for SP and EDF can be interpreted similarly.

Next we describe a schedulability condition for a statistical service which exploits statistical multiplexing gain. The condition requires stationarity and independence of flows. For the purposes of this note, we make the convenient assumption that

$$Pr\left[\sup_\tau \left\{\sum_p A_{\mathcal{C}_p}(t - \tau, t + x_p^{q,\tau}) - \tau\right\} \leq d_q\right]$$

$$\approx \inf_\tau Pr\left[\sum_p A_{\mathcal{C}_p}(t - \tau, t + x_p^{q,\tau}) - \tau \leq d_q\right] . \tag{8}$$

Assuming that eq. (8) holds with equality, we have that an arbitrary type-q arrival has a deadline violation with probability $< \varepsilon$ if d_q is selected such that

$$\sup_\tau \left\{\sum_p \mathcal{G}_{\mathcal{C}_p}(x_p^{q,\tau} + \tau, \varepsilon/Q) - \tau\right\} \leq d_q . \tag{9}$$

Remark: The drawback of the condition in eq. (9) is the dependence on the assumption in eq. (8). Since the assumption does not hold in general, the resulting schedulability condition may be overly optimistic. In Boorstyn et al. [4], it was shown that a more conservative effective envelope, called a "global effective envelope," can result in a rigorous bound which does not require the assumption of eq. (8).

3 EVALUATION OF MULTIPLEXING GAIN

We will now evaluate the multiplexing of MPEG video sources using the schedulability conditions from subsection 2.2. The performance measure for the evaluation is the number of video flows that can be provisioned on a link with delay guarantees. The following allocation methods will be considered in the evaluation:

- *Peak rate allocation*: A peak rate allocation reserves bandwidth at the peak rate of a traffic flow. While a peak rate allocation yields deterministic delay bound guarantees, it does not exploit any multiplexing gain, and, therefore, is an inefficient method for achieving delay guarantees. The number of flows that can be supported with a peak rate allocation serves as a lower bound for any method for provisioning delay guarantees.
- *Deterministic allocation*: Here, we use the schedulability condition from eq. (7) and obtain a service with deterministic delay guarantees. A deterministic allocation captures multiplexing gain achievable through scheduling, but does not exploit statistical multiplexing gain.

TABLE 1   Parameters of the movie traces.

Movie Trace	Average frame size (bits/frame)	Mean Rate (Mbps)	Peak Rate (Mbps)
Terminator	10,904	0.261	1.90
Lambs	7,312	0.171	3.22

- *Statistical allocation*: We use eq. (9) to determine admissibility of flows with the effective envelope from eq. (6). The service guarantees of the statistical allocation are probabilistic delay guarantees. The statistical allocation exploits statistical multiplexing gain, as well as the multiplexing gain due to scheduling.
- *Average rate allocation*: An average rate allocation merely guarantees average throughput and finiteness of delays, but does not support delay guarantees. Since the number of flows admitted with an average rate allocation is always close to 100% of the link capacity, the average rate allocation provides an upper bound for the number of flows admitted by an allocation method.

We use statistics of MPEG-compressed video as traffic sources. The evaluation with MPEG streams is analogous to that in Wrege et al. [60], which explored the multiplexing gain of a service with deterministic delay guarantees. In our examples, a number of MPEG-compressed video sequences are multiplexed on 622-Mbps links. We assume that the video sequences are played continuously with a randomly shifted starting time chosen uniformly over the length of the trace. We consider two traces of MPEG-compressed video from Rose [57]. The first trace is taken from the movie "Terminator 2" (*Terminator*), and the second trace is obtained from the movie "Silence of the Lambs" (*Lambs*). Both traces are digitized to 384 by 288 pixels with 12 bit color information and compressed at 24 frames per second with frame pattern IBBPBBPBBPBB (12 frames). Each sequence consists of a total of 40,000 video frames, corresponding to approximately 30 minutes of video. The data of these traces is given in terms of frame sizes. In table 1, we show some of the parameters of the traces. We assume that the arrival of a frame is spread evenly over an interframe interval (of length 1/24 s); Hence, a (normally instantaneous) frame arrival occurs at a constant rate.

For each of the MPEG traces, we assume that a deterministic regulator is obtained using the method described in Wrege et al. [60]: (1) empirical envelopes are obtained from the MPEG traces using eq. (1), (2) the concave hull of the empirical envelopes is approximated by a piecewise linear function, and (3) the segments of the resulting functions yield a set of rate and burst parameters, which determines a deterministic envelope function as in eq. (2). In table 2 we present the parameters which are obtained from the two MPEG traces with this method.

TABLE 2   Rate and burst parameters of the movie traces using the algorithm from Wrege et al. [60].

Silence of the Lambs (*Lambs*)	
Rate parameter (Bits per second)	Burst parameter t(Bits)
$\rho_1 = 3,221,376.0$	$\sigma_1 = \quad 0.0$
$\rho_2 = \quad 867,008.0$	$\sigma_2 = \quad 98,098.7$
$\rho_3 = \quad 759,628.8$	$\sigma_3 = \quad 156,262.4$
$\rho_4 = \quad 694,336.0$	$\sigma_4 = \quad 246,149.3$
$\rho_5 = \quad 656,472.0$	$\sigma_5 = \quad 321,122.0$
$\rho_6 = \quad 647,850.7$	$\sigma_6 = \quad 372,131.6$
$\rho_7 = \quad 563,438.9$	$\sigma_7 = 1,126,242.3$
$\rho_8 = \quad 502,912.0$	$\sigma_8 = 2,042,261.3$
$\rho_9 = \quad 448,013.1$	$\sigma_9 = 2,911,892.3$
$\rho_{10} = \quad 208,800.0$	$\sigma_{10} = 3,157,800.0$
Terminator 2 (*Terminator*)	
Rate parameter (Bits per second)	Burst parameter (Bits)
$\rho_1 = 1,909,440.0$	$\sigma_1 = \quad 0.0$
$\rho_2 = \quad 869,056.0$	$\sigma_2 = \quad 43,349.3$
$\rho_3 = \quad 791,680.0$	$\sigma_3 = \quad 75,589.3$
$\rho_4 = \quad 624,776.3$	$\sigma_4 = \quad 165,995.4$
$\rho_5 = \quad 592,576.0$	$\sigma_5 = \quad 214,296.0$
$\rho_6 = \quad 425,421.1$	$\sigma_6 = \quad 485,922.6$
$\rho_7 = \quad 361,641.5$	$\sigma_7 = \quad 679,919.0$
$\rho_8 = \quad 346,464.0$	$\sigma_8 = \quad 961,968.0$
$\rho_9 = \quad 317,920.0$	$\sigma_9 = 1,563,770.7$
$\rho_{10} = \quad 304,514.7$	$\sigma_{10} = 1,853,100.7$

3.1   EXAMPLE 1: COMPARISON OF ENVELOPE FUNCTIONS FOR MPEG TRACES

We first compare envelope functions for MPEG traces. A deterministic envelope of a flow j is given by $A_j^*(\tau) = \min_k\{\sigma_{jk} + \rho_{jk}\tau\}$, where the parameters $\{\sigma_{jk}, \rho_{jk}\}_{k=1,\dots,K}$, given in table 2, are obtained from a concave hull of the empirical envelope of the movie traces [60]. The effective envelope for a group of flows is obtained from the deterministic envelopes using eq. (6).

Figures 5(a) and 5(b), respectively, show the results for N multiplexed *Lambs* and *Terminator* traces, where N is set to $N = 100, 1000$, and $10,000$

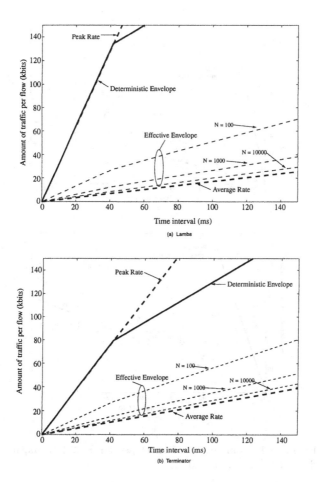

FIGURE 5 Example 1: Comparison of envelopes for $\tau \leq 150\ ms$, $\varepsilon = 10^{-6}$, and for number of flows $N = 100, 1000, 10000$.

flows. In the graphs, we plot the size of the envelopes normalized by the number of flows as functions of time. We use $\varepsilon = 10^{-6}$ for all envelopes.

We observe that the effective envelopes are much smaller than the deterministic envelope or the peak rate. Note that increasing the number of flows N increases the statistical multiplexing gain, leading to a lower traffic rate for each flow.

In figure 6 we show the shape of the envelopes for a fixed number of flows, $N = 1000$, and different values of ε, namely $\varepsilon = 10^{-3}, 10^{-6}$, and 10^{-9}. Figure 6 shows that the effective envelopes are not very sensitive to variations of the parameter ε.

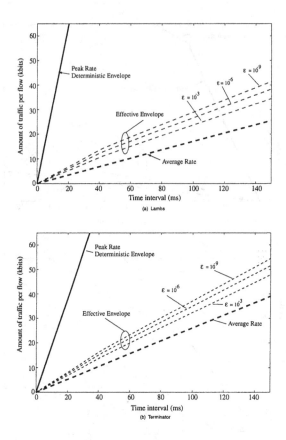

FIGURE 6 Example 1: Comparison of envelopes for $\tau \leq 150 \ ms$, number of flows $N = 1000$ and $\varepsilon = 10^{-3}$, 10^{-6}, and 10^{-9}.

In figure 7 we show how the effective envelopes vary if the number of flows N is increased. We consider the values of the envelopes at the (fixed) time interval $\tau = 50$ ms. For comparison, we include the peak and average rates into the graph. When N is large, the effective envelopes are close to the average traffic rate, indicating a significant statistical multiplexing gain.

3.2   EXAMPLE 2: NUMBER OF ADMITTED MPEG FLOWS

We consider a single link with an FCFS scheduler and compare how many flows can be admitted without violation of deterministic or probabilistic delay guarantees. We also include results from simulations of the statistical rate allocation. The traffic sources are either all flows from the *Lamb* MPEG trace or all flows from the *Terminator* MPEG trace. The results for this example are shown in

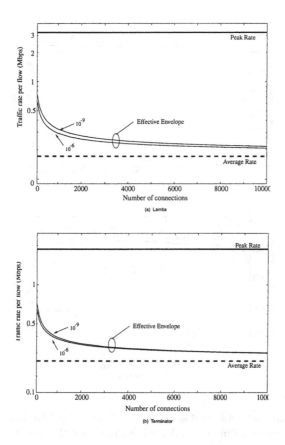

FIGURE 7 Example 1: Traffic rates $\mathcal{G}_C(\tau;\varepsilon)/(N\tau)$ of *Lambs* and *Terminator* for $\tau =$ 50 ms and $\varepsilon = 10^{-6}$ or 10^{-9}.

figure 8. In the example, we set $C = 622$ Mbps and $\varepsilon = 10^{-6}$. The figure shows the number of admitted flows as a function of the delay bound. The results in figure 8 show that a deterministic allocation improves upon a peak rate allocation. However, the statistical allocation, which expresses the statistical multiplexing gain, admits almost as many flows as an average rate allocation scheme.

In figure 9, we show how the achievable average utilization of a link with an FCFS multiplexer increases as the capacity of the link is increased. We fix the delay bound of traffic to $d = 50$ ms and we set $\varepsilon = 10^{-6}$. Figure 9 illustrates the achievable average link utilization as a function of the link capacity. The average achievable link utilization is the sum of the average rates of flows which are admitted according to a chosen allocation method.

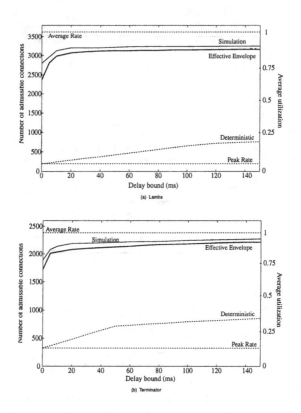

FIGURE 8   Example 2: Admissible number of flows at an FCFS scheduler for flows from the same type as a function of delay bounds, $C = 622$ Mbps, $\varepsilon = 10^{-6}$.

The results in figures 9(a) and (b) show that for both MPEG traces, an average utilization of more than 80% is attainable if the link capacity is above 600 Mbps or more.

3.3   EXAMPLE 3: NUMBER OF ADMITTED MPEG FLOWS WITH DIFFERENT TYPES

Finally, we explore the multiplexing gain at a link with two types of traffic, flows of type *Lambs* and flows of type *Terminator*. The link has a capacity of 622 Mbps. The delay bounds are set to $d_{\text{Terminator}} = 50$ ms for flows of type *Terminator*, and to $d_{\text{Lambs}} = 100$ ms for flows of type *Lambs*.

We consider two scheduling algorithms: Static Priority (SP) and Earliest-Deadline-First (EDF). For purposes of comparison, we include results for a peak rate allocation, average rate allocation, and deterministic delay guarantees.

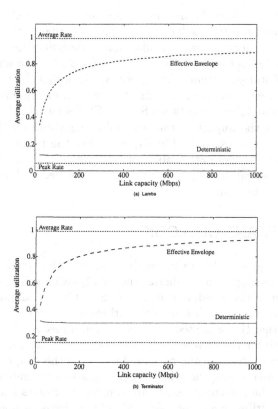

FIGURE 9   Example 2: Average utilization vs. link capacity, $\varepsilon = 10^{-6}$ and $d = 50$ ms.

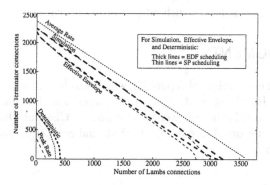

FIGURE 10   Example 3: Admissible region of multiplexing *Lambs* and *Terminator* flows with $\varepsilon = 10^{-6}$ and $d_{\text{Terminator}} = 50$ ms and $d_{\text{Lambs}} = 100$ ms.

In figure 10, we show the maximum number of admitted MPEG flows for the various allocation methods. The figure illustrates that a statistical allocation method admits significantly more traffic than a deterministic allocation. Since the difference between the deterministic and statistical allocation method is the consideration of statistical multiplexing, we conclude that the multiplexing gain due to statistical multiplexing is significant. For both deterministic and statistical allocations, the difference between SP and EDF schedulers is modest. Hence, we conclude that the impact of the scheduling algorithm on the multiplexing gain is limited in this example. Finally, note that the results for the effective envelope are close to those attainable with an average rate allocation. This indicates that statistical delay guarantees can be provided without leaving many network resources unused.

4   SUMMARY AND OUTLOOK

The main observation in the examples is that, at high enough data rates, the statistical multiplexing gain dominates the multiplexing gain, and, in comparison, the selection of the scheduling algorithm has only a small impact. Thus, the use of sophisticated packet scheduling algorithms may have only limited benefits. Note, however, that the presented analysis as limitations. First, the assumptions which were made for the analysis, e.g., the stationarity for MPEG traffic, may not hold in practice. Second, the presented results in this chapter are specific to the employed video traces, the selected delay guarantees, and the selected link capacity. Third, the presented results are single-node results and are not easily generalized to multi-node environments. Finally, we have assumed that the network performs admission control and resource reservation for each video flow. Since per-flow signaling and resource reservation incur significant overhead and are believed to limit the scalability of video networks, it is desirable to analyze multiplexing gain in a network without per-flow reservation.

ACKNOWLEDGMENTS

This note is based on a technical report [3], which was jointly written with Robert Boorstyn, Almut Burchard, and Chaiwat Oottamakorn. The numerical computations presented in this chapter were performed by Chaiwat Oottamakorn.

This work is supported in part by the National Science Foundation through grant ANI-0085955.

REFERENCES

[1] Anick, D., D. Mitra, and M. Sondhi. "Stochastic Theory of a Data-Handling System with Multiple Sources." *Bell Syst. Tech. J.* **61(8)** (1982): 1871–1894.

[2] ATM Forum. Traffic Management, version 4.0, Approved Specifications. April 1996. ⟨http://www.atmforum.com/standards/approved.html⟩.

[3] Boorstyn, R., A. Burchard, J. Liebeherr, and C. Oottamakorn. "Statistical Multiplexing Gain of Link Scheduling Algorithms in QoS Networks." CS-99-21, Computer Science Department, University of Virginia, Charlottesville, VA, 1999

[4] Boorstyn, R., A. Burchard, J. Liebeherr, and C. Oottamakorn. "Statistical Service Assurances for Traffic Scheduling Algorithms." *IEEE J. Selected Areas in Commun.* Special Issue on Internet QoS. **18(12)** (2000): 2651–2664.

[5] Braden, R., D. Clark, and S. Shenker. "Integrated Services in the Internet Architecture: An Overview." RFC 1533, RFC Index, Internet RFC/STD/FYI/BCP Archives, July, 1994.

[6] Burchard, A., Liebeherr, J., and S. D. Patek. "A Calculus for End-to-end Statistical Service Guarantees." CS-01-19, Computer Science Department, University of Virginia, May, 2002.

[7] Castillo, E. *Extreme Value Theory in Engineering.* Academic Press, 1988.

[8] Chang, C. "Stability, Queue Length, and Delay of Deterministic and Stochastic Queueing Networks." *IEEE Trans. Auto. Control* **39(5)** (1994): 913–931.

[9] Choe, J., and Ness B. Shroff. "A Central-Limit-Theorem-Based Approach for Analyizing Queue Behavior in High-Speed Networks." *IEEE/ACM Trans. Network.* **6(5)** (1998): 659–671.

[10] Clark, D., S. Shenker, and L. Zhang. "Supporting Real-Time Applications in an Integrated Services Packet Network: Architecture and Mechanism." In *Proceedings of ACM SIGCOMM'92*, 14–26. New York: ACM Press, 1992.

[11] Cruz, R. L. "A Calculus for Network Delay, Part I: Network Elements in Isolation." *IEEE Transaction of Information Theory* **37(1)** (1991): 114–121.

[12] Cruz, R. L. "A Calculus for Network Delay, Part II: Network Analysis." *IEEE Trans. Info. Theor.* **37(1)** (1991): 132–141.

[13] Daigle, J. N. *Queueing for Telecommunications.* Reading, MA: Addison-Wesley, 1992.

[14] Doshi, B. T. "Deterministic Rule Based Traffic Descriptors for Broadband ISDN: Worst Case Behavior and Connection Acceptance Control." In *The Fundamental Role of Teletraffic in the Evolution of Telecommunications Networks*, edited by J. Labetoulle and J. W. Roberts, 591–600. Elsevier Science B.V., 1994.

[15] Duffield, N. G., and S. H. Low. "The Cost of Quality in Networks of Aggregate Traffic." In *Proceedings of IEEE INFOCOM'98*, 525–532. Washington, DC: IEEE Press, 1998.

[16] Elwalid, A., and D. Mitra. "Design of Generalized Processor Sharing Schedulers Which Statistically Multiplex Heterogeneous QoS Classes." In *Proceedings of IEEE INFOCOM'99*, 1220–1230. Washington, DC: IEEE Press, 1999.

[17] Elwalid, A., and D. Mitra. "Effective Bandwidth of General Markovian Traffic Sources and Admission Control of High Speed Networks." *IEEE/ACM Trans. Network.* **1(3)** (1993): 329–343.

[18] Elwalid, A., D. Mitra, and R. Wentworth. "A New Approach for Allocating Buffers and Bandwidth to Heterogeneous, Regulated Traffic in an ATM Node." *IEEE J. Sel. Areas Comm.* **13(6)** (1995): 1115–1127.

[19] Feng, W.-C., and J. Rexford. "Performance Evaluation of Smoothing Algorithms for Transmitting Prerecorded Variable-Bit-Rate Video." *IEEE Trans. Multimedia* **1(3)** (1999): 302–313.

[20] Ferrari, D., and D. Verma. "A Scheme for Real-Time Channel Establishment in Wide-Area Networks." *IEEE J. Sel. Areas Comm.* **8(3)** (1990): 368–379.

[21] Garrett, M. W., and W. Willinger. "Analysis, Modeling and Generation of Self-Similar VBR Video Traffic." In *Proc. ACM SIGCOMM '94*, 269–280. New York: ACM Press, 1994.

[22] Georgiadis, L., R. Guérin, and V. Peris. "Efficient Network QoS Provisioning Based on per Node Traffic Shaping." *IEEE/ACM Trans. Network.* **4(4)** (1996): 482–501.

[23] Grossglauser, M., S. Keshav, and David N. C. Tse. "RCBR: A Simple and Efficient Service for Multiple Time-Scale Traffic." *IEEE/ACM Trans. Network.* **5(6)** (1997): 741–755.

[24] Guerin, R., H. Ahmadi, and M. Naghshineh. "Equivalent Capacity and Its Application to Bandwidth Allocation in High-Speed Networks." *IEEE J. Sel. Areas Comm.* **9(7)** (1991): 968–981.

[25] Guillemin, F., C. Rosenberg, and J. Mignault. "On Characterizing an ATM Source via the Sustainable Cell Rate Traffic Descriptor." In *Proc. IEEE INFOCOM '95*, 1129–1136. Washington, DC: IEEE Press, 1995.

[26] Hyman, J. M., A. A. Lazar, and G. Pacifici. "Real-Time Scheduling with Quality of Service Constraints." *IEEE J. Sel. Areas Comm.* **9(7)** (1991): 1052–1063.

[27] Jelenkovic, P. R., A. A. Lazar, and N. Semret. "The Effect of Multiple Time Scales and Subexponentiality in MPEG Video Streams on Queueing Behavior." *IEEE J. Sel. Areas Comm.* Special Issue on Real-Time Video Services in Multimedia Networks. **15(6)** (1997): 1052–1071.

[28] Kelly, F. "Notes on Effective Bandwidths." In *Stochastic Networks: Theory and Applications*, edited by F. P. Kelly, S. Zachary, and I. B. Ziedins, 141–168. New York: Oxford University Press, 1996.

[29] Kesidis, G., and T. Konstantopoulos. "Extremal Shape-Controlled Traffic Patterns in High-Speed Networks." *IEEE Trans. Comm.* **48(5)** (2000): 813–819. A shorter version appeared in *Proc. IEEE CDC*, Tampa, FL, Dec. 1998.

[30] Kesidis, G., and T. Konstantopoulos. "Extremal Traffic and Worst-Case Performance for Queues with Shaped Arrivals." In *Analysis of Communication Networks: Call Centres, Traffic and Performance*, edited by D. R. McDonald and S. R. E. Turner. Fields Institute Communications/AMS,

2000. Originally presented at the Fields Institute, U. Toronto, Nov. 9-13, 1998.

[31] Kesidis, G., J. Walrand, and C. Chang. "Effective Bandwidths for Multiclass Markov Fluids and Other ATM Sources." *IEEE/ACM Trans. Network.* **1(4)** (1993): 424–428.

[32] Knightly, E. "H-BIND: A New Approach to Providing Statistical Performance Guarantees to VBR Traffic." In *Proceedings of IEEE INFOCOM'96*, 1091–1099. Washington, DC: IEEE Press, 1996.

[33] Knightly, E. "Enforceable Quality of Service Guarantees for Bursty Traffic Streams." In *Proceedings of IEEE INFOCOM'98*, 635–642. Washington, DC: IEEE Press, 1998.

[34] Knightly, E. W., and P. Rossaro. "Effects of Smoothing on End-to-End Performance Guarantees for VBR Video." In *Proc. International Symposium on Multimedia Communications and Video.* Paper presented at the International Symposium on Multimedia Communications and Video Coding, Brooklyn, NY, October, 1995.

[35] Knightly, E. W., and Ness B. Shroff. "Admission Control for Statistical QoS: Theory and Practice." *IEEE Network* **13(2)** (1999): 20–29.

[36] Knightly, E., and H. Zhang. "Traffic Characterization and Switch Utilization Using Deterministic Bounding Interval Dependent Traffic Models." In *Proc. IEEE INFOCOM '95*, 1137–1145. Washington, DC: IEEE Press, 1995.

[37] Krunz, M., and S. K. Tripathi. "Impact of Video Scheduling on Bandwidth Allocation for Multiplexes MPEG Streams." *Multimedia Systems* **5(6)** (1997): 347–357.

[38] Kurose, J. "On Computing Per-session Performance Bounds in High-Speed Multi-hop Computer Networks." *Proceedings of the 1992 ACM SIGMETRICS Joint International Conference on Measurement and Modeling of Computer Systems*, 128–139. New York: ACM Press, 1992.

[39] Lee, T., K. Lai, and S. Duann. "Design of a Real-Time Call Admission Controller for ATM Networks." *IEEE/ACM Trans. Network.* **4(5)** (1996): 758–765.

[40] Leland, W. E., M. S. Taqqu, W. Willinger, and D. V. Wilson. "On the Self-Similar Nature of Ethernet Traffic (Extended Version)." *IEEE/ACM Trans. Network.* **2(1)** (1994): 1–15.

[41] Liebeherr, J., and D. E. Wrege. "An Efficient Solution to Traffic Characterization of VBR Video in Quality-of-Service Networks." *ACM/Springer Multimedia Sys. J.* **6(4)** (1998): 271–284.

[42] Liebeherr, J., D. Wrege, and D. Ferrari. "Exact Admission Control For Networks With Bounded Delay Services." *IEEE/ACM Trans. Network.* **4(6)** (1996): 885–901.

[43] Liebeherr, J., S. D. Patek, and E. Yilmaz. "Tradeoffs in Designing Networks with End-to-End Statistical QoS Guarantees." *Telecommunication Systems* **23(1-2)** (2003): 9–34.

[44] LoPresti, F., Z. Zhang, D. Towsley and J. Kurose. "Source Time Scale and Optimal Buffer/Bandwidth Tradeoff for Regulated Traffic in an ATM Node." In *Proceedings of IEEE INFOCOM'97*, 676–683. Washington, DC: IEEE Press, 1997.

[45] Oechslin, P. "On-Off Sources and Worst Case Arrival Patterns of the Leaky Bucket." UCL Computer Science research note 97/125, Department of Computer Science, University College of London, UK, 1997.

[46] Pancha, P., and M. El Zarki. "Leaky Bucket Access Control for VBR MPEG Video." In *Proc. IEEE INFOCOM '95*, 796–803. Washington, DC: IEEE Press, 1995.

[47] Papoulis, A. *Probability, Random Variables, and Stochastic Processes.* 3rd ed. McGraw Hill, 1991.

[48] Parekh, A. K., and R. G. Gallager. "A Generalized Processor Sharing Approach to Flow Control in Integrated Services Networks. *IEEE/ACM Trans. Network.* **1(3)** (1993): 344–357.

[49] Parekh, A. K., and R. G. Gallager. "A Generalized Processor Sharing Approach to Flow Control—The Single Node Case." *IEEE/ACM Trans. Network.* **1(3)** (1993): 344–357.

[50] Parekh, A. K., and R. G. Gallager. "A Generalized Processor Sharing Approach to Flow Control in Integrated Services Networks: The Multiple Node Case." *IEEE/ACM Trans. Network.* **2(2)** (1994): 37–150.

[51] Rajagopal, S., M. Reisslein and K. W. Ross. "Packet Multiplexers with Adversarial Regulated Traffic." In *Proceedings of IEEE INFOCOM'98*, 347–355. San Francisco, 1998.

[52] Rathgeb, E. P. "Policing of Realistic VBR Video Traffic in an ATM Network." *Intl. J. Digital and Analog Commun. Syst.* **6** (1993): 213–226.

[53] Reibman, A. R., and A. W. Berger. "Traffic Descriptors for VBR Video Teleconferencing Over ATM Networks." *IEEE/ACM Trans. Network.* **3(3)** (1995): 329–339.

[54] Reisslein, M., K. W. Ross and S. Rajagopal. "Guaranteeing Statistical QoS to Regulated Traffic: The Multiple Node Case." In *Proceedings of 37th IEEE Conference on Decision and Control (CDC)*, 531–538. Washington, DC: IEEE Press, 1998.

[55] Reisslein, M., K. W. Ross, and S. Rajagopal. "Guaranteeing Statistical QoS to Regulated Traffic: The Single Node Case." In *Proceedings of EEE INFO-COM'99*, 1061–1072. Washington, DC: IEEE Press, 1999.

[56] Roberts, J., U. Mocci, and J. Virtamo, eds. *Broadband Network Traffic: Performance Evaluation and Design of Broadband Multiservice Networks.* Final Report of Action. COST 242. Lecture Notes in Computer Science. Vol. 1152. Berlin: Springer-Verlag, 1996.

[57] Rose, O. "Statistical Properties of MPEG Video Traffic and Their Impact on Traffic Modeling in ATM Systems." In *Proceedings of the 20th Annual IEEE Conference on Local Computer Networks*, 397. Washington, DC: IEEE Press, 1995.

[58] Schwartz, M. *Broadband Integrated Networks.* Prentice Hall, 1996.

[59] Starobinski, D., and M. Sidi. "Stochastically Bounded Burstiness for Communication Networks." In *Proc. IEEE INFOCOM '99*, 36–42. Washington, DC: IEEE Press, 1999v.

[60] Wrege, D., E. Knightly, H. Zhang, and J. Liebeherr. "Deterministic Delay Bounds for VBR Video in Packet-Switching Networks: Fundamental Limits and Practical Tradeoffs." *IEEE/ACM Trans. Network.* **4(3)** (1996): 352–362.

[61] Yaron, O., and M. Sidi. "Performance and Stability of Communication Networks via Robust Exponential Bounds." *IEEE/ACM Trans. Network.* **1(3)** (1993): 372–385.

[62] Yates, D., J. Kurose, D. Towsley, and M. Hluchyi. "On Per-Session End-to-End Delay Distributions and the Call Admission Problem for Real-Time Applications with QOS Requirements." In *Proceeedings of ACM SIGCOMM'93*, 2–12. New York: ACM Press, 1993.

[63] Zhang, H. "Providing End-to-End Performance Guarantees Using Non-Work-Conserving Disciplines." *Computing Comunications* **18(10)** (1995): 769–781.

[64] Zhang, H., and D. Ferrari. "Improving Utilization for Deterministic Service In Multimedia Communication." In *Proceedings of the IEEE International Conference on Multimedia Computing and Systems*, 295–304. Washington, DC: IEEE Press, 1994.

[65] Zhang, H., and D. Ferrari. "Rate-Controlled Service Disciplines." *J. High Speed Networks* **3(4)** (1994): 389–412.

[66] Zhang, H., and E. Knightly. "Providing End-to-End Statistical Performance Guarantees with Bounding Interval Dependent Stochastic Models." In *Proceedings of ACM SIGMETRICS'94*, 211–220. New York: ACM Press, 1994.

[67] Zhang, Z., J. F. Kurose, J. D. Salehi, and D. F. Towsley. "Smoothing, Statistical Multiplexing, and Call Admission Control for Stored Video." *IEEE J. Sel. Areas Comm.* Special Issue on Real-Time Video Services in Multimedia Networks. **15(6)** (1997): 1148–1166.

Content Networks:
Taxonomy and New Approaches

H. T. Kung
C. H. Wu

In this chapter we describe a taxonomy for content networks and suggest new architectures for such networks. In recent years, many types of content networks have been developed in various contexts, including peer-to-peer networks [8, 12, 16, 27, 31, 36, 49], cooperative Web caching [3, 47], content distribution networks [2, 7, 11], subscribe-publish networks [6, 52], and content-based sensor networks [14, 21, 22]. For each context, there have been numerous architectural approaches with various design objectives. Our taxonomy attempts to formulate a design space for both existing and future content networks and to identify design points of interest. The proposed new content networks, called semantic content-sensitive networks, offer desirable features such as support for content-proximity searches and the use of small routing tables.

The Internet as a Large-Scale Complex System,
edited by Kihong Park and Walter Willinger, Oxford University Press.

1   INTRODUCTION

A content network is an overlay Internet Protocol (IP) network [4, 38] that supports content routing; that is, messages are routed on the basis of their contents rather than on the IP address of their destinations. Permanent binding of contents to hosts will no longer be necessary to provide access to the contents. Nodes of the overlay network, called network or content nodes, route messages and may also store contents. By using overlay networks, content networks have the flexibility to customize their topologies to meet specific application needs and performance objectives [37, 45, 60].

Content networks can be attractive for a number of reasons. At any given time, a piece of content and its copy may be freely placed at or moved to any network node to improve content availability, minimize access time, support source anonymity, etc. Because content routing reflects the content of a message, a content network allows properly provisioned links to facilitate the routing of contents that belong to classes deemed valuable. In addition, security can be based on content rather than on site [8, 52, 53].

In this chapter we describe a taxonomy for content networks, illustrate various designs in the taxonomy with existing or proposed content networks, indicate strengths of different design approaches, and introduce a new class of content networks that perform "semantic aggregation and content-sensitive placement" of contents. This class corresponds to a design point in our taxonomy that has not been not sufficiently exploited in the literature even though, as we point out, it enjoys certain desirable features, such as the support of semantic proximity searching, not shared by other architectures.

This chapter has the following outline. Section 2 describes a classification scheme for content networks based on two dimensions: content aggregation and content placement. Section 3 discusses the properties and presents examples of the four types of content networks in the proposed taxonomy. Sections 4 and 5 give examples of the new type of content networks that use semantic aggregation and content-sensitive placement of content. Section 6 illustrates that content networks that perform semantic aggregation can use both content-sensitive and content-oblivious placement of contents, thus offering the strengths of both schemes. Finally, section 7 summarizes and concludes the chapter.

2   TWO CLASSIFYING DIMENSIONS FOR CONTENT NETWORKS

We classify content networks on the basis of their attributes in two dimensions:

1. Content aggregation: semantic vs. syntactic, and
2. Content placement: content-sensitive vs. content-oblivious.

In section 3.2 these dimensions are used to classify many existing or proposed content networks.

Dimension (1) relates to the use of content aggregation for the purposes of placement and routing. These functions can be performed either on a content-group or on an individual content basis. As described below, using content aggregation, a content network can scale up to a massive amount of content. Dimension (2) relates to the placement of contents on network nodes. The placement strategy affects optimization of content routing and the size of routing tables. These two dimensions are orthogonal, in that design choices in one do not necessarily dictate those in the other.

2.1 CONTENT AGGREGATION: SEMANTIC VS. SYNTACTIC

Content aggregation is a process of assigning individual contents to content groups. The process can be considered to consist of two steps. The first, "aggregation mapping," maps individual contents to values in some value space. The second, "aggregation grouping," groups individual contents on the basis of those mapped values.

Content aggregation can be "semantic" or "syntactic," and content networks that use semantic or syntactic aggregation are called semantic or syntactic content networks, respectively.

A. Semantic Aggregation. For semantic aggregation, individual contents are mapped to values that have "meaning" with respect to some external system, and the contents are grouped according to their mapped values in the context of that external system. Consider, for example, an aggregation that uses an animal taxonomy as its external system. In the first aggregation step, contents related to cats and dogs are mapped to the values "cat" and "dog," respectively. In the second step, because cats and dogs are both mammals, the contents are grouped together as the mammal aggregate.

When semantic aggregation is used, the contents grouped in the same aggregate all satisfy certain common features. For example, the mammal aggregate contains contents related to animals that share the biological feature of breast-feeding. Thus, contents in the same aggregate may be characterized, related, or compared in terms of shared features in the associated external system. An interesting consequence is that a semantic content network can allow semantic-approximate or proximity searching [10, 17, 25]. That is, contents that are semantically related to a given piece of content can be found by searching through network nodes that contain an aggregate to which that content belongs. For example, a search through the mammal aggregate will find contents related not only to dogs and cats but also to mice and elephants, and so on

B. Syntactic Aggregation. For syntactic aggregation, the mapped values of contents do not belong to any external system of interest. Consider, for example, an aggregation mapping that is a hash function to be applied to a syntactic identifier of a piece of content, such as its file name. In this case, the mapped value of a piece of content is just a bit-string, which does not reveal the meaning of the content in any useful way. From the bit-string value resulting from a hash calculation, one cannot tell the nature of the content. Furthermore, there is no a priori external system in which comparing bit-string values would make sense. Nevertheless, in this case, one can still group contents on the basis of their hashed bit-string values, as in some peer-to-peer networks [8, 12, 27]. But contents in the same content aggregate will not share any common features or meaning derived from any external system of interest to applications. This type of aggregation is thus called syntactic, because it lacks semantic meaning.

Because the contents of the same aggregate are not related according to any external criteria of interest, content grouping based on syntactic aggregation can improve the scalability of a content network without revealing the nature of the contents in a group [12, 27, 36, 49]. Syntactic aggregation can be useful when content anonymity is a design objective [8].

Note that a content network may not perform any content aggregation at all; in other words, content placement and routing are both performed on an individual-content basis. We use the convention that content networks that perform no aggregation are considered syntactic networks.

2.2   CONTENT PLACEMENT: CONTENT-SENSITIVE VS. CONTENT-OBLIVIOUS

The placement of content in a content network can be either "content-sensitive" or "content-oblivious." That is, the location where a piece of content or a content group is placed can be either a function of the content or independent of it.

A. Content-Sensitive Placement.   For content-sensitive placement, the location of a piece of content or a content group in a network is a function of the content. Both semantic and syntactic content networks can employ content-sensitive placement. Consider, for instance, a semantic content network with a tree topology that aggregates contents based on an animal taxonomy. Suppose that content placement follows the rule that contents related to dogs can be placed only in a certain subtree reserved for dog-related contents. This, then, is an example of a semantic content network that uses content-sensitive placement. See sections 4 and 6 for detailed examples of semantic content networks using content-sensitive placement of content.

Consider next an example of a syntactic content network with a hypercube topology. Suppose that contents are aggregated on the basis of hashed values of their file names and that the placement of the content also is based on these values. Thus, all the contents with the same hashed values are placed at the same network node [12, 27]. This is an example of a syntactic content network that uses content-sensitive placement.

With content-sensitive placement, content routing can be made to satisfy specific properties. Consider again the syntactic content network with the hypercube topology, mentioned above. Content-sensitive placement can ensure that, given the hashed values of a piece of content, that content can always be reached from any node through a fixed route determined by those values [12, 27].

For semantic content networks, content-sensitive placement can be especially useful. For example, sports contents can be placed on a sports subnet, basketball contents on a subnet of the sports subnet, NBA contents on a subnet of the basketball subnet, etc. In this sense, the topology hierarchy matches the content hierarchy.

Matching topology and content hierarchy yields a number of advantages, some of which are listed here:

- Efficient processing of search queries. For example, queries for content related to the NBA can first be forwarded over the links leading to the sports subnet, then those to the basketball subnet, and, finally, the NBA subnet.
- Small routing tables. The content-routing table of a node can be as small as the number of content groups at the corresponding content layer, independent of the total number of contents in these groups. Consider, for example, a network node that forwards queries over the link that leads to the basketball subnet. Its content-routing table will need to contain only a few dozen types of sports, such as basketball, football, tennis, golf, etc.
- Semantic-approximate or proximity search. For example, while at a network node that contains basketball-related contents, a search process can consult all these contents beyond those of the NBA, such as information about other basketball leagues.
- Properly provisioned network paths for content access. Because the route to any given content from the root of the topology hierarchy is fixed, the underlying path can be properly provisioned to facilitate access to the content. For example, if contents related to the NBA are popular, then high-bandwidth pipes or shortcut links (see section 4.4) may be allocated between the sports subnet and the NBA subnet. This is in contrast to today's Internet, where bandwidth or link provision for content access can be difficult. For example, in an autonomous system of the Internet, routes are computed using cost metrics based on hop counts with no concern for content.

In content-sensitive placement, the original content may be mirrored at a number of network nodes to facilitate access to the content by nearby nodes [12, 27]. Content-sensitive placement refers to the restriction that the placement of the original contents needs to be content-sensitive. This restriction does not necessarily apply to copies of an item of content, which may be stored at mirror sites or on cached servers.

B. Content-Oblivious Placement.   For content-oblivious placement, content groups can be placed anywhere without consideration of the content. Because the content can be at any place, the network will need either to learn or to find routes to reach them. To support this, the network may rely on a centralized server to maintain locations of content. In this case, network nodes possessing the content may register their content and locations to the server. A node requesting the content may first query the server for their locations, and then send the request to the locations [31].

Several decentralized schemes may also be used here. For example, nodes that contain contents or their copies can periodically advertise the possession of these contents to peering nodes [3, 34]. On the basis of such advertisements, the peering nodes build or update their content routing tables for reaching these contents and also advertise their availability in being used by other nodes to reach the contents. The same procedure is repeated at all nodes, when they receive such advertisements from their peers. Many variations on these schemes are possible, including, for example, schemes based on summary vectors, such as the Bloom filters used in cooperative Web caching [3]. Note that when the contents are network IP addresses, the basic scheme is the traditional distance vector algorithm used in IP routing [34].

Alternatively, nodes that request contents can advertise their queries to their peers. Upon receiving a query, a node can reply that it possesses the requested content or, if it does not, relay the request to its peers [8, 16].

In content-oblivious placement, a content network allows unconstrained placement of contents on network nodes with no regard for the contents or for the network topology. This convenience is at the expense of transmitting content advertisements or query messages and requiring relatively large routing tables. That is, content-oblivious placement exchanges routing overheads and routing table size for convenient network topology management and content placement.

3   TAXONOMY FOR CONTENT NETWORKS

There are four types of content networks based on the two-dimensional classification scheme (see section 2)]. For a summary, see table 1. Each type corresponds to one of the following four policies regarding treatment of content:

A. Syntactic aggregation and content-oblivious placement;
B. Syntactic aggregation and content-sensitive placement;
C. Semantic aggregation and content-oblivious placement; and
D. Semantic aggregation and content-sensitive placement.

TABLE 1   Four types of content networks based on the classifying dimensions in section 2.

	Syntactic aggregation of content	Semantic aggregation of content
Content-oblivious placement of content	Type A: Syntactic content-oblivious network	Type C: Semantic content-oblivious network
Content-sensitive placement of content	Type B: Syntactic content-sensitive network	Type D: Semantic content-sensitive network

3.1   PROPERTIES OF CONTENT NETWORKS

From our discussion in section 2, we can infer properties for each type of content network. For example, content networks that perform syntactic aggregation of their contents can naturally support content anonymity, because no semantic information about the contents is used. Content networks that perform semantic aggregation of their contents can facilitate content proximity searches, because contents with similar meaning are placed together in the network. Content networks that use content-sensitive placement of their contents can support a massive amount of individual contents. This is because the size of the routing tables will depend on the number of content groups a node handles, independent of the total amount of content in the network. Finally, content networks that use content-oblivious placement can be highly fault-tolerant, because the contents and their copies can be placed anywhere and routes to their current locations are learned dynamically.

3.2   EXAMPLES OF CONTENT NETWORKS

It appears that most of the content networks now deployed belong to types A, B, and C. Type D has not received much attention in the literature. This section provides examples of content networks of all four types. Additional examples of type D content networks are described in sections 4 and 5.

Type A Syntactic Content-Oblivious Network.   A simple example of this type of content network is one related to Web proxy servers [3, 47]. In this case, the contents are URLs, and network nodes are a set of Web proxy servers. Because no content aggregation is performed, this is a syntactic content network. Furthermore, because any given content can reside at any of the proxy servers, it is a content network that uses content-oblivious placement. A similar example is a set of mirror sites that host specific contents [2, 7, 11].

Many existing content-delivering systems that rely on centralized servers to locate contents without aggregating them can be viewed as syntactic content-oblivious networks. In such systems, the contents can be placed on any node disregarding their meaning, and their locations are registered in a centralized server. These systems include search engines, such as Google [18] and Yahoo [59], where Web pages are contents hosted on Web servers linked by the URLs with which the search engines respond to queries. A set of Web services [51, 54], described with WSDL [58], communicated with SOAP [57], and registered at the UDDI servers [50], may also be viewed as syntactic content-oblivious networks.

Most existing peer-to-peer (P2P) systems that do not aggregate contents, such as Napster, Gnutella, and Freenet, are syntactic content-oblivious networks. The contents in these systems may be placed on any node, independent of the location of other contents. The contents are searched via centralized servers (e.g., Napster [31, 46]) or through query-flooding (e.g., Gnutella [9, 16] and Freenet [8, 15]). These systems usually incur serious scalability problems [1, 26, 40, 48].

Consider P2P systems that use query flooding. In the case of Gnutella, when a node receives a query that it cannot serve owing to the absence of the requested content from the local store, the node will forward the query to other nodes in a breadth-first manner [9, 16]. A node will stop forwarding a query that was previously received or whose time-to-live counter has expired. In contrast, in the case of Freenet, depth-first search is used [8, 15]. When receiving a query from its parent peer node, a node will serve the request if it has the requested content in the local store. Otherwise, it will forward the query to the one child that can most likely serve the query. If this child fails to serve the query, then the node will try another child with the next highest potential for success in serving the query. If none of the children succeeds, the node will notify its parent peer node that it cannot serve the request.

SIENA [6], a subscribe-publish network that provides event notification services, is another example of syntactic content-oblivious networks. In SIENA, the client, called a subscriber or publisher, issues a subscription to or a notification of events to a nearby server, and a set of interconnected servers that advertise and multicast the events to interested parties. In this case, the contents are subscriptions and notifications, and the network nodes are subscribers, publishers, and forwarding servers. A notification is pushed to interested subscribers that have previously advertised related subscriptions. To match notifications and subscriptions, a network node may use a filter or pattern to match a set of events. Because content is not aggregated and identical or similar events could be issued by any subscriber or publisher, SIENA is a syntactic content-oblivious network. Similarly, Publius [52], a system for Web publishing, is a syntactic content-oblivious network.

The subscribe-publish model has been used in sensor networks such as SCADDS [14, 21] and SPIN [22]. These are syntactic content-oblivious networks, in which the contents are detected data or signals and the content nodes are sensors.

Type B Syntactic Content-Sensitive Network. Some P2P systems that support distributed hashing tables [37] have adopted syntactic aggregation and content-sensitive placement strategies in order to avoid using nonscalable schemes, such as centralized servers or flooding query, described above. Several examples are mentioned here.

PAST is a persistent P2P storage utility that uses a hypercube-based routing scheme, called Pastry, to route contents [12, 41, 42]. In Pastry, every peer node is assigned a 128-bit node identifier (NodeId), derived from a cryptographic hash of the node's public key. Every file is assigned a quasi-unique 160-bit file identifier (FileId) that corresponds to the cryptographic hash of the file's textual name, the owner's public key, and a random salt. Nodes in Pastry are connected using a hypercube topology, in which adjacent nodes share some common address prefixes. Pastry routes a message concerning a given FileId toward the node whose NodeId is numerically closest, among all live nodes, to the 128 most significant bits of the FileId. This means that contents are aggregated syntactically according to the FileId and placed at the numerically closest node.

Like Pastry, OceanStore [27] uses a hypercube algorithm, called Tapestry [62], which is a variation on a randomized hierarchical distributed data structure [35] that can deterministically locate an object. In Tapestry, every server is assigned a random (and unique) NodeId, and every object is identified by a globally unique identifier (GUID), which is a secure hash of the owner's key and some human-readable name. Objects are syntactically aggregated and content-sensitively placed in a manner similar to the process in Pastry. OceanStore is a highly redundant, persistent storage system. Before invoking Tapestry, OceanStore will first use a probabilistic algorithm called attenuated Bloom filters to track and locate objects. Each node in OceanStore maintains an array of Bloom filters where the ith Bloom filter is the union of the Bloom filters for all the nodes at distance i from the current node. A query is routed to the node whose filter indicates the presence of the object at the smallest distance.

Unlike these examples that use hypercube structures, Chord architecture [49] uses a ring topology to provide scalable peer-to-peer lookup services. Nodes and files are assigned m-bit identifiers computed by a hash function. The nodes form an identifier circle in the order of the identifiers' numeric values. Given a key such as the hashed identifier of a file name, Chord maps the key onto a node. The file with FileId k will then be assigned to the first node with an identifier equal to or that follows k in the identifier space. For the node with NodeId i and its preceding node with NodeId j in the identifier circle, files with FileId k where $j < k \leq i$ are aggregated and placed at the node with NodeId j. CAN [36] uses a virtual d-dimensional coordinate space, which may be viewed as a generalization of the one-dimensional identifier scheme above. At any time the entire coordinate space is dynamically partitioned among all the nodes such that each node in CAN owns an individual, distinct zone in the overall space. Every file is represented as a point in a specific zone maintained by some node. A CAN node holds information about a small number of adjacent zones in the space. Using a simple greedy forwarding scheme, the node routes a message to the adjacent zone with coordinates closest to the destination coordinate.

Type C Semantic Content-Oblivious Network.   In semantic content-oblivious networks, the contents are aggregated semantically and the resultant content groups can be placed anywhere. Query broadcast or source advertisement is used to search the contents (see section 2.2B). TRIAD [19] is an example of this type of content network. The contents are mapped to their URLs and grouped on the basis of the domain name portion of the URL. Thus, contents with the same domain name are aggregated into the same group. The URL structure is the "external system" as discussed in section 2.1A. In TRIAD, the Internet Name Resolution Protocol (INRP) performs name lookup, whereas the Name-Based Routing Protocol (NBRP) performs routing advertisement. Like BGP [39], NBRP allows content aggregations to reside anywhere in the content network. In addition, TRIAD proposes to use routing aggregates to reduce the size of name-routing tables.

Type D Semantic Content-Sensitive Network.   An example of a semantic content-sensitive network is the well-known Internet Domain Name System (DNS) [28, 29, 30]. Consider the DNS service that translates host names or domain names into their IP addresses. In this case, the contents are host names and domain names described in fully qualified domain names. In the DNS lookup service, the contents are semantically aggregated into groups according to existing organizational hierarchies, and the content aggregates are placed at DNS servers that belong to the corresponding organizations. Thus, the DNS network is a content network that performs semantic aggregation and content-sensitive placement of content. The next two sections give further examples of content networks of this type, which use data-driven and static content grouping.

4   A SEMANTIC CONTENT-SENSITIVE NETWORK: USING A DATA-DRIVEN CONTENT GROUPING

This section describes a semantic content-sensitive network in which the aggregation grouping is data-driven (see section 2.1), in the sense that it will result from a clustering analysis of all the contents present in the network. First, the data-driven grouping is described, and then the construction of the corresponding semantic content-sensitive network.

Note that in section 5, the same method for constructing a content network is used for the case of static grouping of content.

4.1   DATA-DRIVEN CONTENT GROUPING

This section describes a clustering method for grouping a given set of contents using a clustering tree derived from content vectors associated with the contents.

First, the content vectors are described, then the clustering tree, and, finally, the grouping method.

A. Content Vectors. We assume that the given contents are associated with content vectors that indicate similarity or dissimilarity. The content vector associated with a piece of content can be considered its "mapped value" in the aggregation process (see section 2.1). For example, when the content is a Web page, the associated content vector may be expressed in terms of the frequencies of certain key words in the page. Such vectors are similar to term weighting in the vector space model used in conventional information retrieval systems [44]. For some typical methods of assigning content vectors to contents, refer to Salton and Buckley [43]. In the rest of section 4, we use the terms "content" and the associated "content vector" interchangeably.

There are two major categories of similarity metrics for content vectors [25]. These are angle-based metrics which use, for example, the cosine function, and distance-based metrics which use the inner-product function. For the discussion below, we assume distance-based metrics.

Figure 1 illustrates a set of two-dimensional content vectors associated with eight pieces of content: LA, NY, NCAA, PGA, Mozart, Bach, Jack, and Helen. These vectors correspond to points in the two-dimensional space. With a distance-based metric the positions of the content vectors in the two-dimensional space reflect their similarity, that is, contents of similar nature are close to each other. For example, LA and NY are similar in the sense that both are cities that host NBA teams, and thus their vectors in the two-dimensional space are close together. It should be clear that the concepts illustrated here generalize to content vectors of dimensions higher than two.

B. Clustering Trees. Given a set of contents, we construct a clustering tree for their content vectors. A number of methods are available in the literature for the construction of clustering trees [20, 23, 24]. Figure 1 illustrates one of the simplest methods. Using this method, we start by finding a pair of points, P_1 and P_2, that is one of the closest pairs among all pairs of points. Then the two points are connected, and for the rest of the clustering process, they are replaced by a new point P_{12}, which is a midpoint of the segment (P_1, P_2). The process is repeated until all points are connected.

The binary tree that results from the above process of connecting all the content vectors is called a clustering tree. Each of its subtrees defines a cluster that contains the leaf content vectors of the subtree. We call the root of the subtree the centroid of the cluster. Note that in the construction of the clustering tree, each step can be viewed as connecting two centroids and creating a new one. The distance between the two centroids to be connected at each step is nondecreasing as the construction proceeds. Thus, sibling centroids become farther apart or remain at the same distance at the next higher layer of the tree.

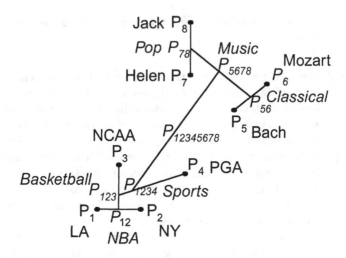

FIGURE 1   Eight content vectors, P_1, P_2, ..., P_8, and their content clustering tree.

A cluster can be represented by its centroid in the sense that the content vectors associated with the cluster are generally closer to this centroid than to its sibling centroid. For example, P_{5678} is the centroid of the cluster consisting of P_5, P_6, P_7, and P_8, and these points are closer to P_{5678} than the sibling P_{1234}.

To reflect the volume of contents in the associated cluster, a centroid might be weighted by moving it toward the centroid of a subtree that has more leaves than the other subtree. For example, in figure 1 the position of a midpoint reflects the "weights" of the two subtrees, that is, the amount of content under them. For example, midpoint P_{123} is closer to P_{12} than to P_3 by a factor of two, because the subtree rooted at P_{12} has twice as many content nodes as the subtree rooted at P_3.

C. Content Grouping Based on A Content Clustering Tree.
Given a content clustering tree, we can aggregate contents in groups, each of which is associated with a centroid. For each internal node of the clustering tree, we use a dividing hyperplane that is perpendicular to the line connecting to the two centroids of the two subtrees [5, 32]. The hyperplane partitions the space of the content vectors into two groups, each containing one of the centroids.

To illustrate the grouping, consider the clustering tree in figure 2(a), where the dividing hyperplanes are denoted by dotted lines. As shown, when content vectors are two-dimensional, a dividing hyperplane is a line. Note that there is a hierarchy among groups, in the sense that a group associated with a centroid of a tree contains the two subgroups associated with the centroids of the two subtrees.

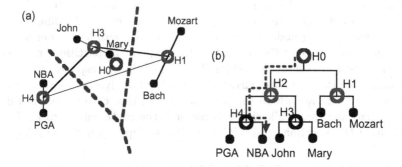

FIGURE 2 (a) Content clustering tree with dividing hyperplanes constructed using Voronoi diagrams; and (b) content networking using the same tree topology and content-sensitive placement of content.

A hyperplane can be weighted by moving it toward the centroid of a subtree that has fewer leaves than the other subtree. In figure 2(a) the dividing hyperplane corresponding to the root of the clustering tree illustrates this weighting. Note that the weighting direction used here is opposite to that used in weighted centroids described earlier in section 4.1B. Both weighting heuristics are for the same purpose, that of increasing the amount of content in the same clusters, and will be grouped together by dividing hyperplanes. Other weighting heuristics could also be used to reflect, for example, expected content access frequencies, rather than the number of contents present.

4.2 CONSTRUCTION OF A CONTENT NETWORK

Suppose that we are given a set of contents with an associated clustering tree. We can construct a content network with the same topology, using content-sensitive placement of the content.

A. Topology of a Content Network. Each node in the content network corresponds to a node of the clustering tree. Two content nodes are connected in the content network if, and only if, the two corresponding nodes in the clustering tree have the parent-child relationship. Consider, for example, the clustering tree in figure 2(a). Figure 2(b) depicts the content network with the same topology.

B. Content-Sensitive Placement of Content. Figure 2 illustrates content-sensitive placement. For a given piece of content, we describe the process of finding the node where the content will be placed. Suppose that the given piece of content has its associated content vector at the location "target" shown in figure 2(a). We show how a route leading to the content node NBA, which is the node closest to the "target," is derived. The route will start at the root H0 of the content network. To determine the next hop, we make use of the dividing hyperplane at the corresponding node in the clustering tree. Because the hyperplane partitions the entire space of the content vectors, one of the partitions must contain the given piece of content. Consider the child of the corresponding node in the clustering tree that is in the partition containing the given piece of content. The next hop will be the node in the content network that corresponds to this child in the clustering tree. Following this method, the route will go to H2 from H0, to H4 from H2, and finally to NBA from H4, as depicted in figure 2(b).

4.3   PROPERTIES OF THE RESULTING CONTENT NETWORK

The content network of figure 2(b) is a semantic content-sensitive network because it uses semantic aggregation and content-sensitive placement of content. The network has the advantages of type D content networks (see section 2.2A). It can support content proximity searches while using relatively small routing tables with sizes independent of the total number of contents in the network. For example, the search suffices to go only to the subtree rooted at H4 for contents related to PGA and NBA. The routing table at H1 or H2 need only contain two entries. If contents related to NBA deserve high-bandwidth access, then specially provisioned network paths can be provided from the root to the node NBA.

4.4   SHORTCUT OPTIMIZATION

Shortcut links can be added to provide direct links between content nodes, as illustrated in figure 3. Using the shortcut links from H0 to H3 and H4 and the existing link from H0 to H1, the content node H0 can directly forward a query to H1, H3, and H4, without going through the intermediate content node H2. The search direction at H0 will now be determined by which of the three sites of figure 3(d) has the target content, rather than either of the two sites of figure 3(c). To determine the next hop from H0, two dividing hyperplanes, rather than one, will be needed, which can be computed using Voronoi diagrams [5, 32].

5   SEMANTIC CONTENT-SENSITIVE NETWORK: USING A STATIC CONTENT GROUPING

This section describes a semantic content-sensitive network in which the aggregation grouping referred to in section 2.1 is static, that is, the association of a piece of content with a group is static and independent of other contents in the network. This is in contrast to the data-driven grouping discussed in section 4.

Many existing content-retrieval systems use static hierarchical structures already in place to group the content. These include library catalog systems [13], DNS [28, 29, 30], and LDAP [61]. In addition, directory-based search engines

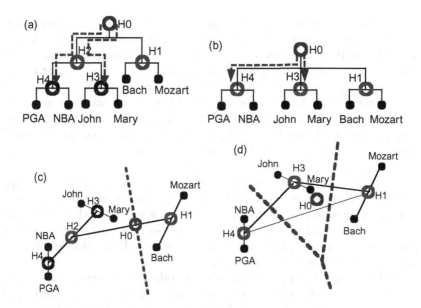

FIGURE 3 Shortcut optimization. (a) Node H0 recognizes that there is heavy traffic from H0 to both H3 and H4, and there is no traffic destined to H2. (b) H0 creates shortcut links to both H3 and H4, without involving H2. H0 will now direct route to one of the three sites of (d) rather than the two sites of (c).

such as Yahoo [59] and Open Directory [33] use catalogs to classify Web links. In all these systems, subject categories form a hierarchical tree, and every object is classified into a subject category. Objects belonging to the same subject category share certain common semantic meanings.

For these systems, semantic content-sensitive networks can be constructed using exactly the same approaches as discussed in section 4, except that the cluster-forming portion of the task is no longer needed. For this reason our description of the construction of the content network here will be brief.

Given a static hierarchical content grouping scheme, we can construct a semantic content-sensitive network. The content network is constructed as described in section 4.2, with the given content hierarchy replacing the clustering tree.

For example, by referring to an external hierarchical taxonomy such as Open Directory [33], we can represent contents and subject categories with XML [55] and RDF [56]. The content network will have the same topology as the taxonomy hierarchy. With contents of the same subject grouped together and placed in the same node, the resultant content network is thus a semantic content-sensitive network.

To process a request related to a given piece of content, we need a method of associating the request with groups that contain the content. One approach, called self-describing requests, is to assume that every request message, such as a content insert or a content query, is self-describing in the sense that the query itself includes specific group information. Thus, when a content node receives such a request, it will know where to forward it on the basis of the group information in the request. This approach is used, for example, by DNS, where every resolving query itself specifies its fully qualified domain name [28, 29, 30]. This allows a node to forward unresolved queries to the proper server responsible for handling domain names in the next level up.

A content network that uses self-describing requests will not directly support a proximity search. Instead, external mechanisms will be used to map a request concerning a piece of content to one concerning a group to which that content belongs. This mechanism is similar to an Internet directory search engine that responds to keyword-based queries with categories or URLs that are deemed to be related to the keywords in the query.

Another approach, called self-classifying nodes, is to assume that every content node is capable of mapping a piece of content related to a request to a proper group to which the content belongs. For example, every content node may keep a copy of a taxonomy database and may determine the proper group for any given piece of content. When the node receives a request message for content query or content insert, it consults its local copy of the taxonomy to determine the next hop to forward the request. For this type of content network, a request message no longer needs to be attached with precise taxonomy information describing the content in the request.

Consider figure 4 for an example of a semantic content-sensitive network that uses self-classifying nodes. Using the content taxonomy in figure 4(a), we construct the content network shown in figure 4(b). In this network, each subject category is assigned to a content node, every node has a copy of the taxonomy of figure 4(a), and the contents are semantically aggregated and placed at proper nodes. When receiving a request to find content near NBA, H0 knows, by consulting its local copy of the taxonomy, that the content belongs to Category Sports (H2), and, similarly, H2 knows it belongs to Category Associations (H4). Because every node has a copy of the taxonomy and needs to have only routing entries to its child nodes, the network can support content proximity searching with relatively small routing tables.

In addition, shortcut optimization, described in section 4.4, can be applied here. For example, if content related to NBA deserves high-bandwidth access, specially provisioned network paths can be supplied from the root to the node NBA, or shortcut links added directly from the root to it.

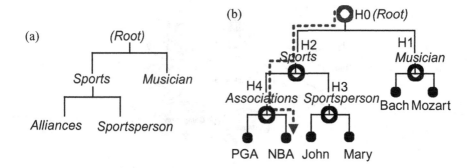

FIGURE 4 (a) A content taxonomy; and (b) content networking using the same tree topology and using content-sensitive placement of content.

6 SEMANTIC CONTENT NETWORK USING HYBRID PLACEMENT

A semantic content network can use content-oblivious and content-sensitive placement for different parts of the network at the same time. This hybrid approach can have the strengths of the content-oblivious scheme such as fault-tolerance and those of the content-sensitive scheme such as small routing tables and content proximity search, as discussed in section 3.1.

Consider, for example, the network model shown in figure 5, where a set of stub networks is connected through a backbone. There are two types of these stub networks: content stub networks and access stub networks. A content stub network is a semantic content network specializing in some content area such as sports, arts, computers, music, or news, using content-sensitive placement. An access stub network is an access network through which users can connect to the backbone. The overall network is semantic in the sense that contents are semantically grouped and placed in the corresponding content stub networks.

Content stub networks use content-sensitive placement of content, which allows the local network administrator to customize content aggregation and placement to support content searches and management. An example of a content stub network can be the network shown in figure 2(b) (see section 4) with its root connected to the backbone.

The backbone uses content-oblivious placement. That is, the root of a content stub network may be placed at any edge node of the backbone, regardless of the content hosted by the content stub network.

The roots of content stub networks will periodically advertise their presence and locations over the backbone, as described in section 2.2B. On the basis of the advertised information, the backbone sets up its content-routing tables.

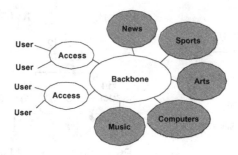

FIGURE 5 A network model consisting of content and access stub networks and a backbone. Content stub networks are denoted by grayed ellipses. The backbone uses content-oblivious placement of content stub networks, whereas content stub networks use content-sensitive placement of contents within their networks.

Using these routing tables, the backbone forwards user requests arriving from access stub networks that address contents in a particular area to the root of the corresponding content stub network.

Alternatively, the roots of content stub networks can register their content and locations in a centralized server, such as a UDDI server [50], with protocols such as SOAP [57]. By querying this server, the background can forward user requests to the roots of those content stub networks which are related to the requested content.

The use of content-oblivious placement over the backbone offers important advantages. Content stub networks can be migrated to or replicated at locations where access to a particular type of content is frequent or has a high value. For example, sports information related to soccer and its mirrors could be kept in a variety of content stub networks situated in Europe, where interest in this sport is enormous. As described above, the backbone will automatically learn the locations of these content stub networks.

The network model shown in figure 5 suggests a deployment strategy for semantic content-sensitive networks. That is, they can initially be deployed as content stub networks.

Finally, we note that the hierarchical network topology of the backbone and stub networks depicted in figure 5 could allow various combinations of content network types. For example, opposite to the scenario described in this section so far, the backbone could use content-sensitive deployment of content stub networks and stub networks could use content-oblivious placement of contents within their networks. This arrangement can make sense when bandwidth is abundant in content stub networks, but it is scarce in the backbone.

7 CONCLUSIONS

Content networks as described in this article are networks in which the addressing and routing of content is based on content rather than on their locations. Thus, in a content network, content is free to move around and be replicated while it remains accessible.

Content networks can address some of the limitations found in current IP networks. For instance, the absence of location addresses on a content network reduces vulnerability to denial of service attacks. Because the location of the content is arbitrary and need not necessarily be associated with a particular server, content providers can be anonymous. Content networks can support new types of client applications that require flexible addressing or no addressing at all, such as peer-to-peer or mobile computing. Content networks can properly provision paths leading to contents deemed to be valuable. As semantic content networks, they support content proximity searches. When content-sensitive placement is used, these networks allow the use of small routing tables to handle massive amounts of content.

In this chapter, we have provided a taxonomy that attempts to capture a design space of both current and future content networks. We have identified design points of interest and characterized their individual strengths. In particular, we have described a new type of semantic content network that uses content-sensitive placement of content and argued that these networks possess desirable properties and deserve further study.

ACKNOWLEDGMENTS

This research was supported in part by the Air Force Office of Scientific Research Multidisciplinary University Research Initiative Grant F49620-97-1-0382, in part by DARPA under Air Force Research Laboratory AFRL/SNKD F33615-01-C-1983, and in parts by research grants from Microsoft Research, Nortel Networks, and Sun Microsystems.

REFERENCES

[1] Adar, E., and B. A. Huberman. "Free Riding on Gnutella." Technical Report, Xerox PARC. August 2000. Available at ⟨http://www.hpl.hp.com/shl/papers/gnutella/Gnutella.pdf⟩.

[2] Akamai Technologies, Inc. Home Page. 2004. ⟨http://www.akamai.com/index_flash.html⟩.

[3] Almeida, J., A. Broder, P. Cao, and L. Fan. "Summary Cache: A Scalable Wide-Area Web Cache Sharing Protocol." *IEEE/ACM Trans. Network.* **8(3)** (2000): 281–293.

[4] Andersen, D., H. Balakrishnan, F. Kaashoek, and R. Morris. "Resilient Overlay Networks." In *Proc. of the 18th ACM Symposium on Operating Systems Principles*. Operating System Review 35(5). New York: ACM Press, 2001.

[5] Aurenhammer, F. "Voronoi Diagram: A Survey of a Fundamental Geometric Data Structure." *ACM Comp. Surveys* **23(3)** (1991): 345–405.

[6] Carzaniga, A., D. Rosenblum, and A. Wolf. "Design and Evaluation of a Wide-Area Event Notification Service." *ACM Trans. Comp. Sys.* **19(3)** (2001): 332–383.

[7] Cisco Systems, Inc. "Application and Content Networking System." Content Delivery Network Solutions. 2003. ⟨http://www.cisco.com/en/US/netsol/ns340/ns394/ns50/ns264/networking_solutions_package.html⟩.

[8] Clarke, I., O. Sandberg, B. Wiley, and T. W. Hong. "Freenet: A Distributed Anonymous Information Storage and Retrieval System." In *International Workshop on Designing Privacy Enhancing Technologies: Design Issues in Anonymity and Unobservability* , 46–66. Lecture Notes in Computer Science, vol. 2009. New York: Springer-Verlag, 2001.

[9] Clip2. "The Gnutella Protocol Specification v0.4." 2002. ⟨http://dss.clip2.com/GnutellaProtocol04.pdf⟩.

[10] Dean, J., and M. Henzinger. "Finding Related Pages in the World Wide Web." In In *Proceedings of WWW-8, the Eighth International World Wide Web Conference*, edited by A. Mendelzon, vol. 31(11–16), 1467–1479. Elsevier Science, 1999.

[11] Digital Island. Home Page. 2002. ⟨http://www.sandpiper.net⟩. Digital Island has been owned by Cable & Wireless since 2001.

[12] Druschel, P., and A. Rowstron. "PAST: A Large-Scale, Persistent Peer-to-Peer Storage Utility." In *Proceedings of the 8th Workshop on Hot Topics in Operating Systems (HotOS-VIII)* , 75. New York: IEEE Press, 2001.

[13] Dublin Core Metadata Initiative. Home Page. August 2004. ⟨http://dublincore.org⟩.

[14] Estrin, D., R. Govindan, J. Heidemann, and S. Kumar. "Next Century Challenges: Scalable Coordination in Sensor Networks." In *Mobile Computing and Networking*, proceedings of ACM MOBICOM'99, 263–270. New York: ACM Press, 1999.

[15] Freenet Project. "Freenet Protocol 1.0 Specification." See "Papers" as no direct link is available. ⟨http://vfreenetproject.org/⟩.

[16] Gnutella Homepage. ⟨http://gnutella.wego.com/⟩.

[17] Goldman, R., N. Shivakumar, S. Venkatasubramanian, and H. Garcia-Molina. "Proximity Search in Databases." In *Proc. of the 24th International Conference on Very Large Data Bases*, 26–37. San Francisco: Morgan Kaufman, 1998.

[18] Google. Home Page. 2004. ⟨http://www.google.com/⟩.

[19] Gritter, M., and D. R. Cheriton. "An Architecture for Content Routing Support in the Internet." In *Proc. of the 3rd USENIX Symposium on Internet Technologies and Systems*. March 2001. ⟨http://www.usenix.org/events/usits01/technical.html⟩.

[20] Han, E.-H., and G. Karypis. "Centroid-Based Document Classification: Analysis and Experimental Results." In *Principles of Data Mining and Knowledge Discovery*, Proceedings of the 4th European Conference on Principles of Data Mining and Knowledge Discovery, 424–431. London: Springer-Verlag, 2000.

[21] Heidemann, J., F. Silva, C. Intanagonwiwat, R. Govindan, D. Estrin, and D. Ganesan. "Building Efficient Wireless Sensor Networks with Low-Level Naming." In *Proc. of the 18th ACM Symposium on Operating Systems Principles*, 146–159. New York: ACM Press, 2001.

[22] Heinzelman, W., J. Kulik, and H. Balakrishnan. "Adaptive Protocols for Information Dissemination in Wireless Sensor Networks." In *Mobile Computing and Networking*, proceedings of ACM MOBICOM'99, 174–185. New York: ACM Press, 1999v.

[23] Jain, A., and R. Dubes. *Algorithms for Clustering Data*. New Jersey: Prentice-Hall, 1988.

[24] Jain, A., M. Murty, and P. Flynn. "Data Clustering: A Review." *ACM Comp. Surveys* **31(3)** (1999): 264–323.

[25] Jones, W., and G. Furnas. "Pictures of Relevance: A Geometric Analysis of Similarity Measures." *J. Am. Soc. Infor. Sci.* **38(6)** (1987): 420–442.

[26] Jovanovic, M. "Scalability Issues in Large Peer-to-Peer Networks: A Case Study of Gnutella." Technical Report, University of Cincinnati, Cincinnati, OH, 2001. Available at ⟨http://www.ececs.uc.edu/ mjovanov/Research/paper.html⟩.

[27] Kubiatowicz, J., D. Bindel, Y. Chen, S. Czerwinski, P. Eaton, D. Geels, R. Gummadi, S. Rhea, H. Weatherspoon, W. Weimer, C. Wells, and B. Zhao. "OceanStore: An Architecture for Global-Scale Persistent Storage." In *ACM SIGPLAN Notices*, vol. 35(11), 190–201. New York: ACM Press, 2000.

[28] Mockapetris, P. "Domain Names-Concepts and Facilities." Request for Comments 1034, RFC Index, Internet RFC/STD/FYI/BCP Archives, November, 1987.

[29] Mockapetris, P. "Domain Names-Implementation and Specification." Request for Comments 1035, RFC Index, Internet RFC/STD/FYI/BCP Archives, November, 1987.

[30] Mockapetris, P., and K. Dunlap. "Development of the Domain Name System," *ACM SIGCOMM Computer Communication Review* **18(4)** (1988): 123–133.

[31] Napster Home Page. January 2004. ⟨http://www.napster.com/⟩.

[32] Okabe, A., B. Boots, K. Sugihara, S. N. Chiu, and M. Okabe. *Spatial Tessellations: Concepts and Applications of Voronoi Diagrams*, 2d ed. John Wiley and Sons, July 2000.

[33] Netscape. Open Directory Project Home Page. 1998. ⟨http://dmoz.org/⟩.

[34] Perlman, R. *Interconnections: Bridges, Routers, Switches, and Internetworking Protocols*, 2d ed. Reading, MA: Addison-Wesley, 2000.

[35] Plaxton, C., R. Rajaraman, and A. Richa. "Accessing Nearby Copies of Replicated Objects in a Distributed Environment." In *Proceedings of the Ninth Annual ACM Symposium on Parallel Algorithms and Architectures*, 311–320. New York: ACM Press, 1997.

[36] Ratnasamy, S., P. Francis, M. Handley, R. Karp, and S. Shenker. "A Scalable Content-Addressable Network." In *Proceedings of the 2001 Conference on Applications, Technologies, Architectures, and Protocols for Computer Communications*, 161–172. New York: ACM Press, 2001.

[37] Ratnasamy, S., S. Shenker, and I. Stoica. "Routing Algorithms for DHTs: Some Open Questions." In *Proc. of the First International Workshop on Peer-to-Peer Systems*, 45-52. New York: Springer-Verlag, 2002.

[38] Ratnasamy, S., M. Handley, R. Karp, and S. Shenker. "Topologically-Aware Overlay Construction and Server Selection." In *Proceedings of the Twenty-First Annual Joint Conference of the IEEE Computer and Communications Societies (INFOCOM '02)*, vol. 3, 1190–1199. Washington, DC: IEEE Press, 2002.

[39] Rekhter, Y., and T. Li. "A Border Gateway Protocol 4 (BGP-4)." Request for Comments 1771, RFC Index, Internet RFC/STD/FYI/BCP Archives, March, 1995.

[40] Ritter, J. "Why Gnutella Can't Scale. No, Really." January 2001. ⟨http://www.darkridge.com/~jpr5/⟩.

[41] Rowstron, A., and P. Druschel. "Pastry: Scalable, Distributed Object Location and Routing for Large-Scale Peer-to-Peer Systems." In *Proceedings of the IFIP/ACM International Conference on Distributed Systems Platforms Heidelberg*, 329–350. Lecture Notes in Computer Science. London: Springer-Verlag, 2001.

[42] Rowstron, A., and P. Druschel. "Storage Management and Caching in PAST, a Large-Scale, Persistent Peer-to-Peer Storage Utility." In *Proceedings of the Eighteenth ACM Symposium on Operating Systems Principles*, 188–201. New York: ACM Press, 2001.

[43] Salton, G., and C. Buckley. "Term-Weighting Approaches in Automatic Text Retrieval." *Info. Proc. & Mgmt.* **24(5)** (1988): 513–523.

[44] Salton, G., A. Wong, and C. S. Yang. "A Vector Space Model for Automatic Indexing." *Comm. ACM* **18(11)** (1975): 613–620.

[45] Saroiu, S., P. Gummadi, and S. Gribble. "Exploring the Design Space of Distributed and Peer-to-Peer Systems: Comparing the Web, TRIAD, and Chord/CFS." In *Proc. of the First International Workshop on Peer-to-Peer Systems*, 214–224. New York: Springer-Verlag, 2002.

[46] Scholl, D. R. "'Napster Messages." April 2000. ⟨http://opennap.sourceforge.net/napster.txt⟩.

[47] Squid. Web Proxy Cache Project. 2004. ⟨http://www.squid-cache.org/⟩.

[48] Sripanidkulchai, K. "The Popularity of Gnutella Queries and Its Implications on Scalability." January 2004.
⟨http://www.cs.cmu.edu/ kunwadee/research/p2p/paper.html⟩.

[49] Stoica, I., R. Morris, D. Karger, F. Kaashoek, and H. Balakrishnan. "Chord: A Scalable Peer-to-Peer Lookup Service for Internet Applications." In *Proceedings of ACM SIGCOMM'01*, 149–160. New York: ACM Press, 2001.

[50] OASIS. "Universal Description, Discovery, and Integration (UDDI) Specification." July 2002. Available at ⟨http://www.uddi.org/⟩.

[51] Vaughan-Nichols. "Web Services: Beyond the Hype." *IEEE Comp. Mag.* **35(2)** (2002): 18–21.

[52] Waldman, M., A. Rubin, and L. Cranor. "Publius: A Robust, Tamper-Evident, Censorship-Resistant Web Publishing System." In *Proc. of the 9th USENIX Security Symposium*, 59–72. USENIX Assoc., 2000.

[53] Wang, C., A. Carzaniga, D. Evans, and A. Wolf. "Security Issues and Requirements for Internet-Scale Publish-Subscribe Systems." In *Proceedings of the 35th Hawaii International Conference on Systems Sciences* (CD-ROM), January 2002. Computer Society Press, 2002.

[54] Web Services Interoperability Organization. Home Page. 2004.
⟨http://www.ws-i.org/⟩.

[55] World Wide Web Consortium. "Extensible Markup Language (XML) Specifications." 1996. ⟨http://www.w3c.org/XML/⟩.

[56] World Wide Web Consortium. "Resources Description Framework (RDF) Specification." ⟨http://www.w3c.org/RDF/⟩.

[57] World Wide Web Consortium. "Simple Object Access Protocol (SOAP) Specification." February 1999. ⟨http://www.w3c.org/TR/SOAP/⟩.

[58] World Wide Web Consortium. "Web Services Description Language (WSDL) Specification." May 2000. ⟨http://www.w3.org/TR/wsdl⟩.

[59] Yahoo Home Page. 2001. ⟨http://www.yahoo.com/⟩.

[60] Yang, B., and H. Garcia-Molina. "Comparing Hybrid Peer-to-Peer Systems." In *Proc. of the 27th International Conference on Very Large Data Bases*, 561–570. Berlin: Morgan Kaufmann, 2001.

[61] Yeong, W., T. Howes, and S. Kille. "Lightweight Directory Access Protocol." Request for Comments 1777, RFC Index, Internet RFC/STD/FYI/BCP Archives, March, 1995.

[62] Zhao, B., J. Kubiatowicz, and A. Joseph. "Tapestry: An Infrastructure for Fault-Tolerant Wide-Area Location and Routing." Technical Report UCB/CSD-01-1141, U.C. Berkeley, Berkeley, CA, April 2001.

Computation in the Wild

Stephanie Forrest
Justin Balthrop
Matthew Glickman
David Ackley

1 INTRODUCTION

The explosive growth of the Internet has created new opportunities and risks by increasing the number of contacts between separately administered computing resources. Widespread networking and mobility have blurred many traditional computer system distinctions, including those between operating system and application, network and computer, user and administrator, and program and data. An increasing number of executable codes, including applets, agents, viruses, e-mail attachments, and downloadable software, are escaping the confines of their original systems and spreading through communications networks. These programs coexist and coevolve with us in our world, not always to good effect. Our computers are routinely disabled by network-borne infections, our browsers crash

The Internet as a Large-Scale Complex System,
edited by Kihong Park and Walter Willinger, Oxford University Press.   227

due to unforeseen interactions between an applet and a language implementation, and applications are broken by operating system upgrades. We refer to this situation as *computation in the wild*, by which we mean to convey the fact that software is developed, distributed, stored, and executed in rich and dynamic environments populated by other programs and computers, which collectively form a *software ecosystem*. The thesis of this chapter is that networked computer systems can be better understood, controlled, and developed when viewed from the perspective of living systems.

Taking seriously the analogy between computer systems and living systems requires us to rethink several aspects of the computing infrastructure—developing design strategies from biology, constructing software that can survive in the wild, understanding the current software ecosystem, and recognizing that all nontrivial software must evolve. There are deep connections between computation and life, so much so that in some important ways, "living computer systems" are already around us, and moreover, such systems are spreading rapidly and will have major impact on our lives and society in the future.

In this chapter we outline the biological principles we believe to be most relevant to understanding and designing the computational networks of the future. Among the principles of living systems we see as most important to the development of robust software systems are: modularity, autonomy, redundancy, adaptability, distribution, diversity, and use of disposable components. These are not exhaustive, simply the ones that we have found most useful in our own research. We then describe a prototype network intrusion-detection system, known as LISYS, which illustrates many of these principles. Finally, we present experimental data on LISYS' performance in a live network environment.

2　COMPUTATION IN THE WILD

In this section we highlight current challenges facing computer science and suggest that they have arisen because existing technological approaches are inadequate. Today's computers have significant and rapidly increasing commonalities with living systems, those commonalities have predictive power, and the computational principles underlying living systems offer a solid basis for secure, robust, and trustworthy operation in digital ecosystems.

2.1　COMPLEXITY, MODULARITY, AND LINEARITY

Over the past fifty years, the manufactured computer has evolved into a highly complex machine. This complexity is managed by deterministic digital logic, which performs extraordinarily complicated operations with high reliability, using modularity to decompose large elements into smaller components. Methods for decomposing functions and data dominate system design methods, ranging from object-oriented languages to parallel computing. An effective decomposition requires that the components have only limited interactions, which is to say that overall system complexity is reduced when the components are nearly independent [24]. Such well-modularized systems are "linear" in the sense that they obey an analog of the superposition principle in physics, which states that for affine differential equations, the sum of any two solutions is itself a solution (see Forrest [10] for a detailed statement of the connection). We use the term "linear" (and "nonlinear") in this sense—linear systems are decomposable into independent modules with minimal interactions and nonlinear systems are not. In the traditional view, nearly independent components are composed to create or execute a program, to construct a model, or to solve a problem. With the emergence of large-scale networks and distributed computing, and the concomitant increase in aggregate complexity, largely the same design principles have been applied. Although most single computers are designed to avoid component failures, nontrivial computer networks must also survive them, and that difference has important consequences for design, as a much greater burden of autonomy and self reliance is placed on individual components within a system.

The Achilles heel of any modularization lies in the interactions between the elements, where the independence of the components, and thus the linearity of the overall system, typically breaks down. Such interactions range from the use of public interfaces in object-oriented systems, to symbol tables in compilers, to synchronization methods in parallel processes. Although traditional computing design practice strives to minimize such interactions, those component interactions that are not eliminated are usually assumed to be deterministic, reliable, and trustworthy: A compiler assumes its symbol table is correct; an object in a traditional object-oriented system does not ask where a message came from; a closely coupled parallel process simply waits, indefinitely, for the required response. Thus, even though a traditionally engineered system may possess a highly decoupled and beautifully linear design, at execution time its modules are critically dependent on each other and the overall computation is effectively a monolithic, highly nonlinear entity.

As a result, although software systems today are much larger, perform more functions, and execute in more complex environments compared to a decade ago, they also tend to be less reliable, less predictable, and less secure, because they critically depend on deterministic, reliable and trustworthy interactions while operating in increasingly complex, dynamic, error prone, and threatening environments.

2.2   AUTONOMY, LIVING SOFTWARE, AND REPRODUCTIVE SUCCESS

Eliminating such critical dependencies is difficult. If, for example, a component is designed to receive some numbers and add them up, what else can it possibly do but wait until they arrive? A robust software component, however, could have multiple simultaneous goals, of which "performing an externally assigned task" is only one. For example, while waiting for task input a component might perform garbage collection or other internal housekeeping activities. Designing components that routinely handle multiple, and even conflicting, goals leads to robustness.

This view represents a shift in emphasis about what constitutes a practical computation, away from traditional algorithms which have the goal of finding an answer quickly and stopping, and toward processes such as operating systems that are designed to run indefinitely. In client-server computing, for example, although client programs may start and stop, the server program is designed to run forever, handling requests on demand. And increasingly, many clients are designed for open-ended execution as well, as with web browsers that handle an indefinite stream of page display requests from a user. Peer-to-peer architectures such as the ccr system [1] or Napster [19], which eliminate or blur the client/server distinction, move even farther toward the view of computation as interaction among long-lived and relatively independent processes. Traditional algorithmic computations, such as sorting, commonly exist not as stand-alone terminating computations performed for a human user, but as tools used by higher-level nonterminating computational processes. This change, away from individual terminating computations and toward loosely coupled ecological interactions among open-ended computations, has many consequences for architecture, software design, communication protocols, error recovery, and security.

Error handling provides a clear example of the tension between these two approaches. Viewed algorithmically, once an error occurs there is no point in continuing, because any resulting output would be unreliable. From a living software perspective, however, process termination is the last response to an error, to be avoided at nearly any cost.

As in natural living systems, successful computational systems have a variety of lifestyles other than just the "run once and terminate" algorithms on the one hand and would be immortal operating systems and processes on the other. The highly successful open-source Apache web server provides an example of an alternative strategy. Following the pattern of many Unix daemons, Apache employs a form of queen and workers organization, in which a single long-lived process does nothing except spawn and manage a set of server subprocesses, each of which handles a given number of web page requests and then dies. This hybrid strategy allows Apache to amortize the time required to start each new worker process over a productive lifetime serving many page requests, while at the same time ensuring that the colony as a whole can survive unexpected fatal errors in server subprocesses. Further, by including programmed subprocess death in the architecture, analogous to cellular apoptosis (programmed cell death), Apache deals with many nonfatal diseases as well, such as memory leaks. A related idea from the software fault tolerance community is that of software rejuvenation [22], in which running software is periodically stopped and restarted again after "cleaning" of the internal state. This proactive technique is designed to counteract software aging problems arising from memory allocation, fragmentation, or accumulated state.

In systems such as the Internet, which are large, open, and spontaneously evolving, individual components must be increasingly self-reliant and autonomous. They must also be able to function without depending on a consistent global design to guarantee safe interactions. Individual components must act autonomously, protect themselves, repair themselves, and increasingly, as in the example of Apache, decide when to replicate and when to die. Natural biological systems exhibit organizational hierarchies (e.g., cells, multicellular organisms, ecosystems) similar to computers, but each individual component takes much more responsibility for its own interactions than what we hitherto have required of our computations.

The properties of autonomy, robustness, and security are often grouped together under the term "survivability," the ability to keep functioning, at least to some degree, in the face of extreme systemic insult. To date, improving application robustness is largely the incremental task of growing the set of events the program is expecting—typically in response to failures as they occur. For the future of large-scale networked computation, we need to take a more basic lesson from natural living systems, that survival is more important than the conventional strict definitions of correctness. As in Apache's strategy, there may be leaks, or even outright terminal programming errors, but they won't cause the entire edifice to crash.

Moving beyond basic survival strategies, there is a longer term survivability achieved through the reproductive success of high-fitness individuals in a biological ecosystem. Successful individuals reproduce more frequently, passing on their genes to future generations. Similarly, in our software ecosystem, reproductive success is achieved when software is copied, and evolutionary dead-ends occur when a piece of software fails to be copied—as in components that are replaced (or "patched") to correct errors or extend functionality.

2.3 DISPOSABILITY, ADAPTATION, AND HOMEOSTASIS

As computers become more autonomous, what happens when a computation makes a mistake, or is attacked in a manner that it cannot handle? We advocate building autonomous systems from disposable components, analogous to cells of a body that are replicating and dying continually. This is the engineering goal of avoiding single points of failure, taken to an extreme. Of course, it is difficult to imagine that individual computers in a network represent disposable components; after all, nobody is happy when it's their computer that "dies," for whatever reason, especially in the case of a false alarm. However, if the disposability is at a fine enough grain size, an occasional inappropriate response is less likely to be lethal. Such lower-level disposability is at the heart of Apache's survival strategy, and it is much like building reliable network protocols on top of unreliable packet transfer, as in TCP/IP. We will also see an example of disposability in the following section, when we discuss the continual turnover of detectors in LISYS, known as *rolling coverage*.

Computation today takes place in highly dynamic environments. Nearly every aspect of a software system is likely to change during its normal life cycle, including: who uses the system and for what purpose, the specification of what the system is supposed to do, certain parts of its implementation, the implementors of the system, the system software and libraries on which the computation depends, and the hardware on which the system is deployed. These changes are routine and continual.

Adaptation and homeostasis are two important components of a robust computing system for dynamic environments. Adaptation involves a component modifying its internal rules (often called "learning") to better exploit the current environment. Homeostasis, by contrast, involves a component dynamically adjusting its interactions with the environment to preserve a constant internal state. As a textbook says, homeostasis is "the maintenance of a relatively stable internal physiological environment or internal equilibrium in an organism" [5, p. G–11].

Given the inherent difficulty of predicting the behavior of even static nonlinear systems, and adding the complication of continuously evolving computational environments, the once-plausible notion of finding *a priori* proofs of program correctness is increasingly problematic. Adaptive methods, particularly those used in biological systems, appear to be the most promising near-term approach for modeling inherent nonlinearity and tracking environmental change. The phenomenon of "bit rot" is widespread. When a small piece of a program's context changes, all too often either the user or a system administrator is required to change pathnames, apply patches, install new libraries (or reinstall old ones), etc. Emerging software packaging and installation systems, such as RPM [15] or Debian's `apt-get` utility, provide a small start in this direction. These systems provide common formats for "sporifying" software systems to facilitate mobility and reproduction, and they provide corresponding developmental pathways for adapting the software to new environments. We believe that this sort of adaptability should take place not just at software installation but as an ongoing process of accommodation, occurring automatically and autonomously at many levels, and that programs should have the ability to evolve their functionality over time.

Turning to homeostasis, we see that current computers already have many mechanisms that can be thought of as homeostatic, for example, temperature and power regulation at the hardware level, and virtual memory management and process scheduling at the operating systems level. Although far from perfect, they are essential to the proper functioning of almost any application program. However, mechanisms such as virtual memory management and process scheduling cannot help a machine survive when it is truly stressed. This limitation stems in part from the policy that a kernel should be fair. When processor time is shared among all processes, and memory requests are immediately granted if they can be satisfied, stability cannot be maintained when resource demands become extreme. Fairness necessitates poor performance for all processes, either through "thrashing," when the virtual memory system becomes stressed, or through unresponsiveness, failed program invocations, and entire system crashes, when there are too many processes to be scheduled. One fair response to extreme resource contention, used by AIX [4], is random killing of processes. This strategy, however, is too likely to kill processes that are critical to the functioning of the system. Self-stabilizing algorithms [23] are similar in spirit, leading to systems in which an execution can begin in any arbitrary initial state and be guaranteed to converge to a "legitimate" state in a finite number of steps.

A more direct use of homeostasis in computing is Anil Somayaji's pH system (for process homeostasis) [25, 26]. pH is a Linux kernel extension which monitors every executing process on a computer and responds to anomalies in process behavior, either by delaying or aborting subsequent system calls. Process-level monitoring at the system-call level can detect a wide range of attacks aimed at individual processes, especially server processes such as sendmail and Apache [11]. Such attacks often leave a system functional, but give the intruder privileged access to the system and its protected data. In the human body, the immune system can kill thousands of cells without causing any outward sign of illness; similarly, pH uses execution delays for individual system calls as a way to restore balance. By having the delays be proportional to the number of recent anomalous system calls, a single process with gross abnormalities will effectively be killed (e.g., delayed to the point that native time-out mechanisms are triggered), without interfering with other normally behaving processes. Likewise, small deviations from normal behavior are tolerated in the sense that short delays are transient phenomena which are imperceptible at the user level. pH can successfully detect and stop many intrusions before the target system is compromised. In addition to coping with malicious attacks, pH detects a wide range of internal system-level problems, slowing down the offending program to prevent damage to the rest of the system.

2.4 REDUNDANCY, DIVERSITY, AND THE EVOLUTIONARY MESS

Biological systems use a wide variety of redundant mechanisms to improve reliability and robustness. For example, important cellular processes are governed by molecular cascades with large amounts of redundancy, such that if one pathway is disrupted an alternative one is available to preserve function. Similarly, redundant encodings occur throughout genetics, the triplet coding for amino acids at the nucleotide level providing one example. These kinds of redundancy involve more than the simple strategy of making identical replicas that we typically see in redundant computer systems. Identical replicas can protect against single component failure but not against design flaws. The variations among molecular pathways or genetic encodings thus provide additional protection through the use of diverse solutions.

This diversity is an important source of robustness in biology. As one example, a stable ecosystem contains many different species which occur in highly conserved frequency distributions. If this diversity is lost and a few species become dominant, the ecosystem becomes unstable and susceptible to perturbations such as catastrophic fires, infestations, and disease. Other examples include the variations among individual immune systems in a population and various forms of genetic diversity within a single species. Computers and other engineered artifacts, by contrast, are notable for their lack of diversity. Software and hardware components are replicated for several reasons: economic leverage, consistency of behavior, portability, simplified debugging, and cost of distribution and maintenance. All these advantages of uniformity, however, become potential weaknesses when they replicate errors or can be exploited by an attacker. Once a method is created for penetrating the security of one computer, all computers with the same configuration become similarly vulnerable [13]. Unfortunately, lack of diversity also pervades the software that is intended to defend against attacks, be it a firewall, an encryption algorithm, or a computer virus detector. The potential danger grows with the population of interconnected and homogeneous computers.

Turning to evolution, we see that the history of manufactured computers is a truly evolutionary history, and evolution does not anticipate, it reacts. To the degree that a system is large enough and distributed enough that there is no effective single point of control for the whole system, we must expect evolutionary forces—or "market forces"—to be significant. In the case of computing, this happens both at the technical level through unanticipated uses and interactions of components as technology develops and at the social level from cultural and economic pressures. Having humans in the loop of an evolutionary process, with all their marvelous cognitive and predictive abilities, with all their philosophical ability to frame intentions, does not necessarily change the nature of the evolutionary process. There is much to be gained by recognizing and accepting that our computational systems resemble naturally evolving systems much more closely than they resemble engineered artifacts such as bridges or buildings. Specifically, the strategies that we adopt to understand, control, interact with, and influence the design of computational systems will be different once we understand them as ongoing evolutionary processes.

In this section we described several biological design principles and how they are relevant to computer and software systems. These include modularity, redundancy, distribution, autonomy, adaptability, diversity, and use of disposable components. As computers have become more complex and interconnected, these principles have become more relevant, due to an increasing variety of interactions between software components and a rising number of degrees of freedom for variation in computational environments. Software components are now confronted with challenges increasingly similar to those faced by organic entities situated in a biological ecosystem. An example of how biological organisms have evolved to cope with their environments is the immune system. In the next section we explore a software framework that is specifically patterned after the immune system. In addition to exhibiting many of the general biological design principles advocated above, this framework illustrates a specific set of mechanisms derived from a particular biological system and applied to a real computational problem.

3 AN ILLUSTRATION: LISYS

In the previous section, we outlined an approach to building and designing computer systems that is quite different from that used today. In this section we illustrate how some of these ideas could be implemented using an artificial immune system framework known as ARTIS and its application to the problem of network intrusion detection in a system known as LISYS. ARTIS and LISYS were originally developed by Steven Hofmeyr in his dissertation [16, 17].

The immune system processes peptide patterns using mechanisms that in some cases correspond closely to existing algorithms for processing information (e.g., the genetic algorithm), and it is capable of exquisitely selective and well-coordinated responses, in some cases responding to fewer than ten molecules. Some of the techniques used by the immune system include learning (affinity maturation of B cells, negative selection of B and T cells, and evolved biases in the germline), memory (cross-reactivity and the secondary response), massively parallel and distributed computations with highly dynamic components (on the order of 10^8 different varieties of receptors [27] and 10^7 new lymphocytes produced each day [20]), and the use of combinatorics to address the problem of scarce genetic resources (V-region libraries). Not all of these features are included in ARTIS, although ARTIS could easily be extended to include them (affinity maturation and V-region libraries are the most notable lacunae).

ARTIS is intended to be a somewhat general artificial immune system architecture, which can be applied to multiple application domains. In the interest of brevity and concreteness, we will describe the instantiation of ARTIS in LISYS, focusing primarily on how it illustrates our ideas about computation in the wild. For more details about LISYS as a network intrusion-detection system, the reader is referred to Balthrop et al [2, 3] and Hofmeyr [17].

3.1 THE NETWORK INTRUSION PROBLEM

LISYS is situated in a local-area broadcast network (LAN) and used to protect the LAN from network-based attacks. In contrast with switched networks, broadcast LANs have the convenient property that every location (computer) sees every packet passing through the LAN.[1] In this domain, we are interested in building a model of normal behavior (known as *self* through analogy to the normally occurring peptides in living animals) and detecting deviations from

---

[1]There are several ways in which the architecture could be trivially modified to run in switched networks. For example, some processing could be performed on the switch or router, SYN packets could be distributed from the switch to the individual nodes, or a combination could be tried.

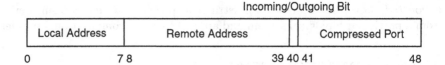

FIGURE 1   The 49-bit compression scheme used by LISYS to represent TCP SYN packets.

normal (known as *non-self*). Self is defined to be the set of normal pairwise TCP/IP connections between computers, and non-self is the set of connections, potentially of enormous size, which are not normally observed on the LAN. Such connections may represent either illicit attacks or significant changes in network configuration or use. A connection can occur between any two computers in the LAN as well as between a computer in the LAN and an external computer. It is defined in terms of its "data-path triple"—the source IP address, the destination IP address, and the port by which the computers communicate [14, 18].

LISYS examines only the headers of IP packets. Moreover, at this point we constrain LISYS to examine only a small set of the features contained in the headers of TCP SYN packets. The connection information is compressed to a single 49-bit string that unambiguously defines the connection. This string is compressed in two ways [17]. First, it is assumed that one of the IP addresses is always internal, so only the final byte of this address needs to be stored. Secondly, the port number is also compressed from 16 bits to 8 bits. This is done by re-mapping the ports into several different classes. Figure 1 shows the 49-bit representation.

3.2   LISYS ARCHITECTURE

LISYS consists of a distributed set of *detectors*, where each detector is a 49-bit string (with the above interpretation) and a small amount of local state. The detectors are distributed in the sense that they reside on multiple hosts; there exists one *detector set* for each computer in the network. Detector sets perform their function independently, with virtually no communication. Distributing detectors makes the system more robust by eliminating the single point of failure associated with centralized monitoring systems. It also makes the system more scalable as the size of the network being protected increases.

A perfect *match* between a detector and a compressed SYN packet means that at each location in the 49-bit string, the symbols are identical. However, in the immune system matching is implemented as binding affinities between molecules, where there are only stronger and weaker affinities, and the concept of perfect matching is ill-defined. To capture this, LISYS uses a partial matching rule known as *r*-contiguous bits matching [21]. Under this rule, two strings match if they are identical in at least r contiguous locations. This means there are $l - r + 1$ windows where a match can occur, where l is the string length. The value r is a threshold which determines the specificity of detectors, in that it controls how many strings can potentially be matched by a single detector. For example, if $r = l$ (49 in the case of LISYS' compressed representation), the match is maximally specific, and a detector can match only a single string—itself. As shown in Esponda [9], the number of strings a detector matches increases exponentially as the value of r decreases.

LISYS uses *negative detection* in the sense that valid detectors are those that fail to match the normally occurring behavior patterns in the network. Detectors are generated randomly and are initially *immature*. Detectors that match connections observed in the network during the *tolerization period* are eliminated. This procedure is known as *negative selection* [12]. The tolerization period lasts for a few days, and detectors that survive this period become *mature*. For the r-contiguous bits matching rule and fixed self sets which don't change over time, the random detector generation algorithm is inefficient—the number of random strings that must be generated and tested is approximately exponential in the size of self. More efficient algorithms based on dynamic programming methods allow us to generate detectors in linear time [6, 7, 28, 29, 30]. However, when generating detectors asynchronously for a dynamic self set, such as the current setting, we have found that random generation works sufficiently well. Negative detection allows LISYS to be distributed, because detectors can make local decisions independently with no communication to other nodes. Negative selection of randomly generated detectors allows LISYS to be adaptive to the current network condition and configuration.

LISYS also uses *activation thresholds*. Each detector must match multiple connections before it is activated. Each detector records the number of times it matches (the *match count*) and raises an alarm only when the match count exceeds the activation threshold. Once a detector has raised an alarm, it returns its match count to zero. This mechanism has a time horizon: Over time the match count slowly returns to zero. Thus, only repeated occurrences of structurally similar and temporally clumped strings will trigger the detection system. The activation threshold mechanism contributes to LISYS's adaptability and autonomy, because it provides a way for LISYS to tolerate small errors that are likely to be false positives.

LISYS uses a "second signal," analogous to costimulatory mechanisms in the immune system. Once a detector is activated, it must be costimulated by a human operator or it will die after a certain time period. Detectors which are costimulated become *memory detectors*. These are long-lived detectors which are more easily activated than non-memory detectors. This secondary response improves detection of true positives, while costimulation helps control false positives.

Detectors in LISYS have a finite lifetime. Detectors can die in several ways. As mentioned before, immature detectors die when they match self, and activated detectors die if they do not receive a costimulatory signal. In addition, detectors

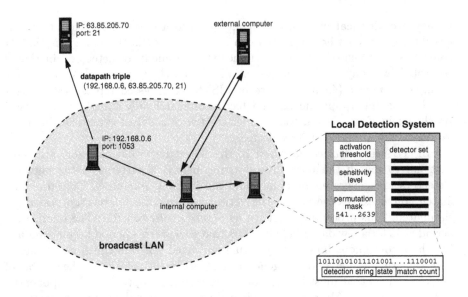

FIGURE 2   The architecture of LISYS.

have a fixed probability of dying randomly on each time step, with the average lifespan of a single detector being roughly one week. Memory detectors are not subject to random death and may thus survive indefinitely. However, once the number of memory detectors exceeds a specified fraction of the total detector population, some are removed to make room for new ones. The finite lifetime of detectors, when combined with detector regeneration and tolerization, results in *rolling coverage* of the self set. This rolling coverage clearly illustrates the principles of disposable components and adaptability.

Each independent detector set (one per host) has a *sensitivity level*, which modifies the local activation threshold. Whenever the match count of a detector in a given set goes from 0 to 1, the effective activation threshold for all the other detectors in the same set is reduced by one. Hence, each different detector that matches for the first time "sensitizes" the detection system, so that all detectors on that machine are more easily activated in future. This mechanism has a time horizon as well; over time, the effective activation threshold gradually returns to its default value. This mechanism corresponds very roughly the effect that inflammation, cytokines, and the other molecules have on the sensitivity of nearbyb individual immune system lymphocytes. Sensitivity levels in LYSYS are a simple adaptive mechanism intended to help detect true positives, especially distributed coordinated attacks.

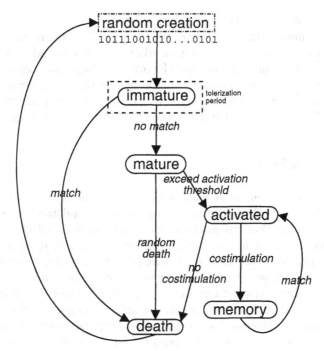

FIGURE 3   The lifecycle of a detector in LISYS.

LISYS uses *permutation masks* to achieve diversity of representation.[2] A permutation mask defines a permutation of the bits in the string representation of the network packets. Each detector set (network host) has a different, randomly generated, permutation mask. One feature of the negative-selection algorithm as originally implemented is that it can result in undetectable patterns called *holes*, or put more positively, *generalizations* [6, 7]. Holes can exist for any symmetric, fixed-probability matching rule, but by using permutation masks we effectively change the match rule on each host, which gives us the ability to control the form and extent of generalization in the vicinity of self [8]. Thus, the permutation mask controls how the network packet is presented to the detection system, which is analogous to the way different MHC types present different sets of peptides on the cell surface.

Although LISYS incorporates many ideas from natural immune systems, it also leaves out many features. One important omission is that LISYS has only a single kind of detector "cell," which represents an amalgamation of T cells, B cells, and antibodies. In natural immune systems, these cells and molecules

---

[2]Permutation masks are one possible means of generating *secondary representations*. A variety of alternative schemes are explored in Hofmeyr [17].

have distinct functions. Other cell types missing from LISYS include effector cells, such as natural killer cells and cytotoxic T cells. The mechanisms modeled in LISYS are almost entirely based on what is known as "adaptive immunity." Also important is innate immunity. Moreover, an important aspect of adaptive immunity is clonal selection and somatic hypermutation, processes which are absent from LISYS. In the future, it will be interesting to see which of these additional features turn out to have useful computational analogs.

4 EXPERIMENTS

In this section we summarize some experiments that were performed in order to study LISYS's performance in a network intrusion-detection setting and to explore how LISYS' various mechanisms contribute to its effectiveness. For the experiments described in this section, we used a simplified form of LISYS in order to study some features more carefully. Specifically, we used a version that does not have sensitivity levels, memory detectors, or costimulation. Although we collected the data on-line in a production distributed environment, we performed our analysis off-line on a single computer. This made it possible to compare performance across many different parameter values. The programs used to generate the results reported in this chapter are available from ⟨http://www.cs.unm.edu/~immsec⟩. The programs are part of LISYS and are found in the LisysSim directory of that package.

4.1 DATA SET

Our data were collected on a small but realistic network. The normal data were collected for two weeks in November 2001 on an internal restricted network of computers in our research group at UNM. The six internal computers in this network connected to the Internet through a single Linux machine acting as a firewall, router and masquerading server. In this environment we were able understand all of the connections, and we could limit attacks. Although small, the network provides a plausible model of a corporate intranet where activity is somewhat restricted and external connections must pass through a firewall. And, it resembles the increasingly common home network that connects to the Internet through a cable or DSL modem and has a single external IP address. After collecting the normal data set, we used a package "Nessus" to generate attacks against the network.

Three groups of attacks were performed. The first attack group included a denial-of-service (DOS) attack from an internal computer to an external computer, attempted exploitations of weaknesses in the configuration of the firewall machine, an attack against FTP (file transfer protocol), probes for vulnerabilities in SSH (secure shell), and probes for services such as chargen and telnet. The second attack group consisted of two TCP port scans, including a stealth scan

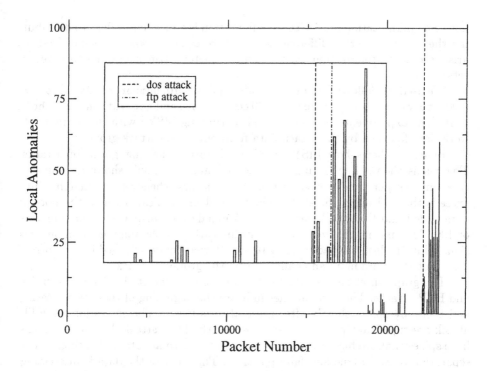

FIGURE 4   Local anomalies for attack group one. The inset is a zoomed view of the same data.

(difficult to detect) and a noisy scan (easy to detect). The third attack group consisted of a full nmap[3] port scan against several internal machines.

Most of these attacks are technically classified as probes because they did not actually execute an attack, simply checking to see if the attack could be performed. However, in order to succeed, a probe would typically occur first; actually executing the attack would create additional TCP traffic, making it easier for LISYS to detect the attack. More details about this data set are given in Balthrop et al. [3].

4.2   DETECTING ABNORMAL BEHAVIOR

Here we report how a simplified version of LISYS performs. In addition to the simplifications mentioned earlier, we also eliminated the use of permutation masks for this set of experiments and used a single large detector set running on a single host. Because we are operating in a broadcast network environment and

---

[3]nmap is a separate software package used by Nessus.

not using permutation masks, this configuration is almost equivalent to distributing the detectors across different nodes (there could be some minor differences arising from the tolerization scheme and parameter settings for different detector sets).

We used the following parameter values [3]: $r = 10$, number of detectors = 6,000, tolerization period = 15,000 connections, and activation threshold = 10. The experiments were performed by running LISYS with the normal network data followed by the attack data from one of the attack groups.

Figure 4 shows how LISYS performed during the first group of attacks. The x-axis shows time (in units of packets) and the y-axis shows the number of anomalies per time period. An anomaly occurs whenever the match count exceeds the activation threshold. The vertical line indicates where the normal data ended and the attack data began. Anomalies are displayed using windows of 100. This means that each bar is an indication of the number of anomalies signaled in the last 100 packets. There are a few anomalies flagged in the normal data, but there are more anomalies during the group of attacks.

The graph inset shows that LISYS was able to detect the denial-of-service and FTP attacks. The vertical lines indicate the beginning of these two attacks, and there are spikes shortly after these attacks began. The spikes for the FTP attack are significantly higher than those for the DOS attack, but both attacks have spikes that are higher than the spikes in the normal data, indicating a clear separation between true and false positives. This view of the data is interesting because the height of the spikes indicates the system's confidence that there is an anomaly occurring at that point in time.

Figure 5 shows the LISYS anomaly data for attack group two. By looking at this figure, we can see that there is something anomalous in the last half of the attack data, but LISYS was unable to detect anomalies during the first half of the attack.[4] Although the spikes are roughly the same height as the spikes in the normal data, the temporal clustering of the spikes is markedly different. Figure 6 shows the LISYS anomaly data for attack group three. The figure shows that LISYS overwhelmingly found the nmap scan to be anomalous. Not only are the majority of the spikes significantly higher than the normal data spikes, but there is a huge number of temporally clustered spikes.

These experiments support the results reported in Hofmeyr [17], suggesting that the LISYS architecture is effective at detecting certain classes of network intrusions. However, as we will see in the next section, LISYS performs much better under slightly different circumstances.

---

[4]In LISYS, detectors are continually being generated, undergoing negative selection, and being added to the repertoire. Some new detectors were added during the first half of the attack, but these detectors turned out not to be crucial for the detection in the second half.

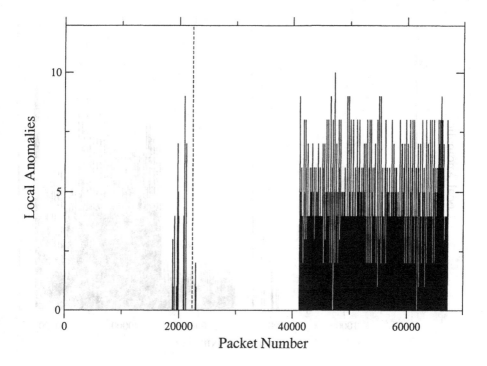

FIGURE 5   Local anomalies for attack group two.

5   THE EFFECT OF DIVERSITY

The goal of the experiments in this section was to assess the effect of diversity in our artificial immune system. Recall that diversity of representation (loosely analogous to MHC diversity in the real immune system) is implemented in LISYS by permutation masks. We were interested to see how LISYS' performance would be affected by adding permutation masks. Because performance is measured in terms of true and false positives, this experiment tested the effect of permutations on the system's ability to generalize (because low false-positive rates correspond to good generalization, and high false positives correspond to poor generalization). For this experiment, 100 sets of detectors were tolerized (generated randomly and compared against self using negative selection) using the first 15,000 packets in the data set (known as the *training set*), and each detector set was assigned a random permutation mask. Each detector set had exactly 5,000 mature detectors at the end of the tolerization period [3], and for consistency we set the random death probability to zero. Five-thousand detectors provides maximal possible coverage (i.e., adding more detectors does not improve subse-

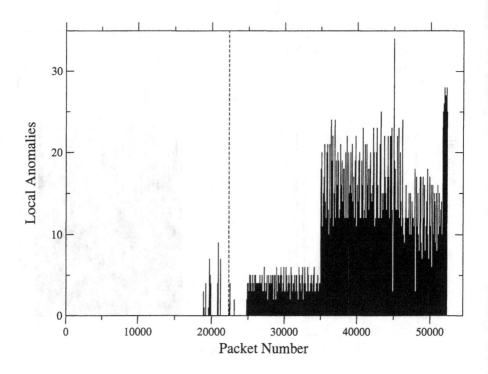

FIGURE 6   Local anomalies for attack group three.

quent matching) for this data set and r threshold. Each set of detectors was then run against the remaining 7,329 normal packets, as well as against the simulated attack data. In these data (the *test sets*), there are a total of 475 unique 49-bit strings. Of these 475, 53 also occur in the training set and are thus undetectable (because any detectors which would match them are eliminated during negative selection). This leaves 422 potentially detectable strings, of which 26 come from the normal set and 396 are from the attack data, making the maximal possible coverage by a detector set 422 unique matches.

An ideal detector set would achieve zero false positives on the normal test data and a high number of true positives on the abnormal data. Because network attacks rarely produce a single anomalous SYN packet, we don't need to achieve perfect true-positive rates at the packet level in order to detect all attacks against the system. For convenience, however, we measure false positives in terms of single packets. Thus, a perfect detector set would match the 396 unique attack strings, and fail to match the 26 new normal strings in the test set.

Figure 7 shows the results of this experiment. The performance of each detector set is shown as a separate point on the graph. Each detector set has its

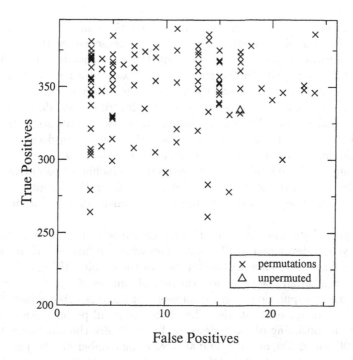

FIGURE 7   LISYS performance under different permutations. Each plotted point corresponds to a different permutation, showing false positives (x-axis) and true positives (y-axis).

own randomly generated permutation of the 49 bits, so each point shows the performance of a different permutation. The numbers on the x-axis correspond to the number of unique self-strings in the test set which are matched by the detector set, i.e., the number of false positives (up to a maximum of 26). The y-axis plots a similar value with respect to the attack data, i.e. the number of unique true positive matches (up to a maximum of 396). The graph shows that there is a large difference in the discrimination ability of different permutations. Points in the upper left of the graph are the most desirable, because they correspond to permutations which minimize the number of false positives and maximize the number of true positives; points toward the lower right corner of the graph indicate higher false positives and/or lower true positives. Surprisingly, the performance of the original (unpermuted) mapping is among the worst we found, suggesting that the results reported in the previous section are a worst case in terms of true vs. false positives. Almost any other random permutation we tried outperformed the original mapping. Although we don't yet have a complete explanation for this behavior, we believe that it arises in the following way.

The LISYS design implicitly assumes that there are certain predictive bit patterns that exhibit regularity in self, and that these can be the basis of distinguishing self from non-self. As it turns out, there are also deceptive bit patterns which exhibit regularity in the training set (observed self), but the regularity does not generalize to the rest of self (the normal part of the test set). These patterns tend to cause false positives when self strings that do not fit the predicted regularity occur. We believe that the identity permutation is bad because the predictive bits are at the ends of the string, while the deceptive region is in the middle. Under such an arrangement, it is difficult to find a window that covers many predictive bit positions without also including deceptive ones. It is highly likely that a random permutation will break up the deceptive region, and bring the predictive bits closer to the middle, where they will appear in more windows.

The preceding analysis is based on the choice of a single permutation. In the original system developed by Hofmeyr, however, each host on the network used a different (randomly chosen) permutation of the 49 bits. Using a diverse set of random permutation masks reduces the overall number of undetectable patterns, thus increasing population-level robustness, but there is a risk of increasing false positives. In the experiments described here, using all permutations (with each permutation consisting of 5,000 detectors) would raise the number of true positives to 396 out of 396, and it would also raise the number of false positives to 26 out of 26 (compared with 334/396 true positives and 17/26 false positives when the original representation is used). However, if we chose only those permutations with low false-positive rates, we can do significantly better. The relative tradeoffs between diversity of representation and the impact on discrimination ability is an important area of future investigation.

6   DISCUSSION

In the previous section we described a prototype network intrusion-detection system based on the architecture and mechanisms of natural immune systems. Although LISYS is not a complete model of the immune system nor a production quality intrusion-detection system, it illustrates many of the principles elucidated in section 2. We identified the principles of redundancy, distribution, autonomy, adaptability, diversity, use of disposable components, and modularity as key features required for successful "computation in the wild."

Of these, LISYS is weakest in the autonomy component. There are a number of ways in which LISYS could be made more autonomous. One way would be to model the so-called constant region of antibody molecules. Detectors in LISYS represent generic immune cells, combining properties of T cells, B cells, and antibodies. However, as described to date they model only the recognition capabilities of immune cells (receptors) and a small amount of state. A simple extension would be to concatenate a few bits to each detector to specify a response (analogous to different antibody *isotypes* which comprise an antibody's "constant region"). This feature would constitute a natural extension to LISYS, although it leaves open the important question of how to interpret the response bits—that is, what responses do we want LISYS to make?

A second approach to adding autonomy to LISYS involves the use of a second signal. Costimulation in LISYS is an example of a second signal, in which a second independent source must confirm an alarm before a response can be initiated. There are several examples of second signals in immunology, including helper T cells (which confirm a B cell's match before the B cell can be activated) and the complement system used in the early immune response. The complement cascade provides a general and early indication of trouble, and operates in combination with highly specific T-cell responses. One possibility for LISYS would be to incorporate the pH system described earlier to provide the second signal. pH can detect anomalous behavior on a host (e.g., during intrusion attempts) and could serve as a generic indicator of "damage" in conjunction with the specific detectors of LISYS. Thus, when a LISYS detector became activated it would respond only if pH confirmed that the system was in an anomalous state. Either of these two approaches would represent an important next step toward autonomy.

7 CONCLUSIONS

The implementation we described is based on similarities between computation and immunology. Yet, there are also major differences, which it is wise to respect. The success of all analogies between computing and living systems ultimately rests on our ability to identify the correct level of abstraction—preserving what is essential from an information-processing perspective and discarding what is not. In the case of immunology, this task is complicated by the fact that natural immune systems handle fundamentally different data streams than those handled by computers. In principle, a computer vision system or a speech-recognition system would take as input the same data as a human does (e.g., photons or sound waves). In contrast, regardless of how successful we are at constructing a computer immune system, we would never expect or want it to handle pathogens in the form of actual living cells, viruses, or parasites. Thus, the level of abstraction for computational immunology is necessarily higher than that for computer vision or speech, and there are more degrees of freedom in selecting a modeling strategy.

Our study of how the architecture of the immune system could be applied to network security problems illustrates many of the important design principles introduced in the first part of this chapter. Many of the principles discussed here are familiar, and in some cases have long been recognized as desirable for computing systems, but in our view, there has been little appreciation of the common underpinnings of these principles, little or no rigorous theory, and few working implementations that take them seriously. Hofmeyr's artificial immune system is, however, an initial step in this direction. As our computers and the software they run become more complex and interconnected, properties such as robustness, flexibility and adaptability, diversity, self reliance and autonomy, and scalability can only become more important to computer design.

ACKNOWLEDGMENTS

The authors gratefully acknowledge the support of the National Science Foundation (grants CDA-9503064, and ANIR-9986555), the Office of Naval Research (grant N00014-99-1-0417), Defense Advanced Projects Agency (grant AGR F30602-00-2-0584), the Santa Fe Institute, and the Intel Corporation.

We thank the many members of Adaptive Computation group at the University of New Mexico for their help over the past ten years in developing these ideas.

REFERENCES

[1] Ackley, D. H. "ccr: A Network of Worlds for Research." In *Artificial Life V. Proceedings of the Fifth International Workshop on the Synthesis and Simulation of Living Systems*, edited by C. G. Langton and K. Shimohara, 116–123. Cambridge, MA: The MIT Press, 1996.

[2] Balthrop, J., F. Esponda, S. Forrest, and M. Glickman. "Coverage and Generalization in an Artificial Immune System." In *GECCO-2002: Proceedings of the Genetic and Evolutionary Computation Conference*, edited by W. B. Langdon, E. Cantu-Paz, K. Mathias, R. Roy, D. Davis, R. Poli, K. Balakrishnan, V. Hanovar, G. Rudolph, J. Wegener, L. Bull, M. A. Potter, A. C. Schultz, J. F. Miller, E. Burke and N. Jonaska, 3–10. New York: Morgan Kaufmann, 2002.

[3] Balthrop, J., S. Forrest, and M. Glickman. "Revisiting Lisys: Parameters and Normal Behavior." In *CEC-2002: Proceedings of the Congress on Evolutionary Computing*. Los Alamitos, CA: IEEE Computer Society Press, 2002.

[4] International Business Machines Corporation. *AIX Version 4.3 System Management Guide: Operating System and Devices*, chapter Understanding Paging Space Allocation Policies. IBM, 1997. ⟨http://www.rs6000.ibm.com/doc_link/en_US/a_doc_lib/aixbman/ baseadmn/pag_space_under.htm⟩.

[5] Curtis, Helena, and N. Sue Barnes. *Biology*. 5th ed. New York: Worth Publishers, Inc., 1989.

[6] D'haeseleer, P., S. Forrest, and P. Helman. "An Immunological Approach to Change Detection: Algorithms, Analysis and Implications." In *Proceed-

ings of the 1996 IEEE Symposium on Computer Security and Privacy. Los Alamitos, CA: IEEE Computer Society Press, 1996.

[7] D'haeseleer, Patrik. "An Immunological Approach to Change Detection: Theoretical Results." In *Proceedings of the 9th IEEE Computer Security Foundations Workshop*. Los Alamitos, CA: IEEE Computer Society Press, 1996.

[8] Esponda, Carlos Fernando, and Stephanie Forrest. "Defining Self: Positive and Negative Detection." In *Proceedings of the IEEE Symposium on Security and Privacy*. Los Alamitos, CA: IEEE Computer Society Press, 2001.

[9] Esponda, Carlos Fernando. "Detector Coverage with the r-Contiguous Bits Matching Rule." Technical Report TR-CS-2002-03, University of New Mexico, Albuquerque, NM, 2002.

[10] Forrest, S. *Emergent Computation*. Cambridge, MA: MIT Press, 1991.

[11] Forrest, S., S. Hofmeyr, A. Somayaji, and T. Longstaff. "A Sense of Self for Unix Processes." In *Proceedings of the 1996 IEEE Symposium on Computer Security and Privacy*. Los Alamitos, CA: IEEE Computer Society Press, 1996.

[12] Forrest, S., A. S. Perelson, L. Allen, and R. Cheru Kuri. "Self-Nonself Discrimination in a Computer." In *Proceedings of the 1994 IEEE Symposium on Research in Security and Privacy*. Los Alamitos, CA: IEEE Computer Society Press, 1994.

[13] Forrest, S., A. Somayaji, and D. H. Ackley. "Building Diverse Computer Systems." In *Sixth Workshop on Hot Topics in Operating Systems*. Los Alamitos, CA: IEEE Computer Society Press, 1998.

[14] Heberlein, L. T., G. V. Dias, K. N. Levitte, B. Mukherjee, J. Wood, and D. Wolber. "A Network Security Monitor." In *Proceedings of the IEEE Symposium on Security and Privacy*. Los Alamitos, CA: IEEE Computer Society Press, 1990.

[15] Herrold, R. P., and the RPM community. "www.rpm.org" (documentation). March 2002. ⟨http:/www.rpm.org/⟩.

[16] Hofmeyr, S. A., and S. Forrest. "Architecture for an Artificial Immune System." *Evol. Comp. J.* **8(4)** (2000): 443–473.

[17] Hofmeyr, Steven A. "An Immunological Model of Distributed Detection and Its Application to Computer Security." Ph.D. thesis, University of New Mexico, Albuquerque, NM, 1999.

[18] Mukherjee, B., L. T. Heberlein, and K. N. Levitt. "Network Intrusion Detection." *IEEE Network* (1994): 26–41.

[19] Napster. March 2002. ⟨http://www.napster.org/⟩.

[20] Osmond, D. G. "The Turn-Over of B-Cell Populations." *Immunology Today* **14(1)** (1993): 34–37.

[21] Percus, J. K., O. Percus, and A. S. Perelson. "Predicting the Size of the Antibody Combining Region from Consideration of Efficient Self/Non-self Discrimination." *PNAS* **90** (1993): 1691–1695.

[22] Pfenning, A., S. Garg, A. Puliafito, M. Telek, and K. Trivedi. "Optimal Software Rejuvenation for Tolerating Software Failures." *Performance Evaluation* **27/28(4)** (1996): 491–506.

[23] Schneider, M. "Self-Stabilization." *ACM Comput. Surveys* **25(1)** (1993): 45–67.

[24] Simon, Herbert A. *The Sciences of the Artificial.* Boston, MA: MIT Press, 1969.

[25] Somayaji, A., and S. Forrest. "Automated Response using System-Call Delays." In *Usenix Security Syposium*, 185–198. New York: ACM Press, 2000.

[26] Somayaji, Anil. "Operating System Stability and Security through Process Homeostasis." Ph.D. thesis, University of New Mexico, Albuquerque, NM, 2002.

[27] Tonegawa, S. "Somatic Generation of Antibody Diversity." *Nature* **302** (1983): 575–581.

[28] Wierzchon, S. T. "Discriminative Power of the Receptors Activated by k-Contiguous Bits Rule." *J. Comp. Sci. & Tech.* **1(3)** (2000): 1–13.

[29] Wierzchon, S. T. "Generating Optimal Repertoire of Antibody Strings in an Artificial Immune System." In *Intelligent Information Systems*, edited by M. A. Klopotek, M. Michalewicz, and S. T. Wierzchon, 119–133. Heidelberg, New York: Springer-Verlag, 2000.

[30] Wierzchon, S. T. "Deriving Concise Description of Non-self Patterns in an Artificial Immune System." In *New Learning Paradigm in Soft Computing*, edited by S. T. Wierzchon, L. C. Jain, and J. Kacprzyk, 438–458. Heidelberg, New York: Springer-Verlag, 2001.

The Bio-Networking Architecture: A Biologically Inspired Approach to the Design of Scalable, Adaptive, and Survivable/Available Network Applications

Tatsuya Suda
Tomoko Itao
Masato Matsuo

In this chapter, the authors propose and describe a new network architecture called the Bio-Networking Architecture. The Bio-Networking Architecture is inspired by the observation that the biological world has already developed mechanisms that are necessary for future network requirements such as self-organization, scalability, adaptation and evolution, security, and survivability. In the biological world, each individual entity (such as a bee in a bee colony) follows a simple set of behavior rules (migration, replication, reproduction, death, energy exchange, and relationship establishment with other entities), yet a group of entities (a bee colony) exhibits complex emergent behavior and characteristics (self-organization, scalability, adaptation and evolution, security and survivability). The authors of this chapter believe that if a network is modeled after biological concepts and mechanisms, it may be able to achieve the desirable properties of self-organization, scalability, adaptation and evolution, security, and survivability.

In the Bio-Networking Architecture, network applications are implemented by a group of distributed, autonomous entities called the cyber-entities (analogous to a bee colony consisting of multiple bees). Each cyber-entity implements a functional component related to its service and also follows simple behavior rules similar to biological entities (such as migration, replication, reproduction, death, energy exchange, and relationship establishment with other cyber-entities). In the Bio-Networking Architecture, useful behaviors and characteristics emerge from the interaction of individual cyber-entities.

There are three innovative features of the Bio-Networking Architecture. First, the Bio-Networking Architecture is the first attempt to apply key biological concepts and mechanisms (such as emergent behavior, adaptation, evolution, diversity, social networking, and energy) to the design of a broad and general class of network applications. Second, the Bio-Networking Architecture enables the construction of complex network applications with the inherent properties of self-organization, scalability, adaptation, evolution, security, and survivability. Third, because the Bio-Networking Architecture adapts and evolves to accommodate short- and long-term changes in network conditions, system designers, administrators, and users are free from managing and tuning network applications.

1   OUR VISION AND BIOLOGICAL INSPIRATIONS

We envision a future where a universal network connects every human being and most human-made electronic objects. In addition to conventional computing devices, most vehicles, appliances, tools, and even unmanned robotic devices in space are nodes in the universal network. Thus, the universal network encompasses locations engaged in every human endeavor, including the home, workplace, cars, trains, airplanes, train stations, airports, and space.

The universal network is viewed as an extension of every single person and organization. The wide availability of open computing resources (openly shared CPU, memory, and bandwidth) and natural interfaces (e.g., speech recognition and brain-computer interfaces) allows the user to merge with the network into a unified entity, giving the person or organization a ubiquitous presence. This unified entity (1) extends the user's identity and presence beyond his or her immediate physical surroundings, (2) extends the user's domain of control beyond his or her physical surroundings, and (3) extends the user's mental capacity beyond human memory and computational abilities.

The unified entity extends a user's identity and presence beyond his or her immediate physical surroundings by maintaining and making available to authorized parties personal information such as medical information, social security number, public encryption key, financial information, personal preferences, and current physical location. Such a network would, for example, allow doctors to instantly access personal medical information. It would allow hotel and airline reservations to be made without specifying preferences (non-smoking, etc.) every time.

The unified entity extends a user's domain of control beyond his or her immediate physical surroundings by allowing immediate control of all devices belonging to the unified entity. For example, a person could be in an airplane and activate the home DVD recorder to record an interesting program that the person just discovered from a friend. For an organization, the unified entity extends the organizational domain of control to remote locations (e.g., the military controlling exploratory sensor vehicles).

Finally, the unified entity extends the user's mental capacity beyond normal human memory and computational abilities by utilizing any available computing resources in the network. Such computing resources provide applications such as personal information management (e.g., calendar, address books), finance (banking, electronic commerce), office productivity (word processors), communication (videoconferencing, collaborative applications), and entertainment (single and multi-player games, movies). Providing access to such essential applications and information from anywhere greatly enhances personal capability and productivity.

Some aspects of this vision are already in progress today (brain-computer interface [55], interplanetary networking [77]). However, we believe a radical shift in network design paradigms is necessary to realize this vision. We believe realizing such a scenario requires a network that exhibits self-organization[1] with inherent support for mobility, scalability, adaptation to short- and long-term changes in network conditions, security, and survivability from massive failures and attacks. Today's networks are unable to satisfactorily meet such challenges.

As inspiration for a new networking paradigm to realize our vision, we observe that the biological world has already developed the mechanisms necessary to achieve some of the desired properties of our envisioned network, such as self-organization, scalability, adaptation and evolution, security and survivability. Many biological systems such as the immune system [5, 68], the bee colony [43, 58], and the ant colony [17, 22] can grow to large populations, adjust to heterogeneous and dynamic environments, detect and eliminate foreign entities, and survive from attacks. For example, a bee colony scales to support a huge number of bees. Bee colonies adapt to a wide variety of weather and food conditions. They secure their hive from intruding bees and other animals, and are able to survive even when a percentage of the bees are destroyed by predators. We believe if networks are modeled after certain biological concepts and mechanisms, they may be able to achieve these desirable properties. Therefore, in this chapter, we propose to apply key concepts and mechanisms from the bi-

---

[1]Self-organization refers to the ability of a system in which the system's components autonomously react to changing environments, interact with one another, and create an ordered system state without requiring central administration or central coordination.

ological world to design a new network architecture called the Bio-Networking Architecture.

2 APPLICATION OF BIOLOGICAL CONCEPTS AND MECHANISMS IN THE BIO-NETWORKING ARCHITECTURE

Some of the biological concepts and mechanisms that we apply in the Bio-Networking Architecture are explained below. Additional biological concepts and mechanisms considered in the Bio-Networking Architecture are described in section 6.

2.1 EMERGENT BEHAVIOR

A key concept in biological systems is *emergent behavior*. In biological systems, useful behavior often emerges through the collective, simple and autonomous behaviors of individual biological entities. For example, when a bee colony needs more food, a large number of bees will leave the hive and go to the flower patches in the area to gather nectar. When the bee colony is near its food storage capacity, only a few bees will leave the hive to gather nectar. This adaptive food gathering function emerges from the relatively simple and local interactions among individual bees. If a returning food gathering bee can quickly unload its nectar to a food storing bee, it means that the food storing bees are not busy and that there is little food in the hive. This food gathering bee then encourages other nearby bees to leave the hive and collect nectar by doing the well-known "waggle dance." If a returning food gathering bee must wait a long time to unload its nectar, it means that the food storing bees are busy and that there is plenty of food in the hive. This food gathering bee then remains in the hive and rests [57, 58]. Since this interaction is localized and only takes place between a food gathering bee and a nearby food storing bee, it can be repeated a number of times to support the large or growing nutritional needs of a bee colony. The bee colony also exhibits other types of emergent behavior such as self-organization, evolution, and survivability. Thus, emergent behavior is the formation of complex behaviors or characteristics through the collective, simple and autonomous behaviors or characteristics of individual entities [2, 17, 63].

In the Bio-Networking Architecture, we apply the concept of emergent behavior by implementing network applications in a group of distributed, self-organizing, self-optimizing, autonomous entities called the *cyber-entities* (CEs) (see fig. 1). This is analogous to a bee colony (an application) consisting of multiple bees (cyber-entities). Each cyber-entity (CE) implements a functional component related to the application and also follows simple behavior rules similar to biological entities (such as reproduction, death, migration, and relationship establishment with other cyber-entities). (Each application compo-

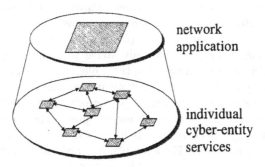

network
application

individual
cyber-entity
services

FIGURE 1   Cyber-entities and network applications.

nent that a cyber-entity implements is referred to as a service in the rest of the chapter.) In the Bio-Networking Architecture, useful application functionality emerges from the collaborative execution of application components (or services) carried by cyber-entities, and useful system behaviors and characteristics (e.g., self-organization, adaptation, evolution, security and survivability) arise from the simple behaviors and interaction of individual cyber-entities.

In the Bio-Networking Architecture, a network application is created from a group of autonomous cyber-entities. The network may provide multiple applications, each of which is implemented with its own group of cyber-entities. Also, a cyber-entity may belong to multiple applications.

2.2   EVOLUTION THROUGH DIVERSITY AND NATURAL SELECTION

Another key concept that we borrow from biological systems is *evolution*. Evolution in biological systems occurs as a result of the natural selection from diverse behavioral characteristics of individual biological entities. Through many successive generations, beneficial features are retained while detrimental behaviors become dormant or extinct, and the biological system specializes and improves itself according to long-term environmental changes.

Within a biological system, *diversity* of behaviors is necessary for the system to adapt and evolve to suit a wide variety of environmental conditions. In the Bio-Networking Architecture, we may design multiple cyber-entities to perform the same task with different behaviors to ensure a sufficient degree of diversity. For instance, we may design multiple cyber-entities to implement the same service with different behavior policies (e.g., one cyber-entity with a migration policy of moving toward a user, and another cyber-entity with a migration policy of moving toward a cheap resource cost node). In the Bio-Networking Architecture, we design cyber-entity reproduction to include mutation and crossover mechanisms to produce new behaviors and new behavior policies.

Within a biological system, food (or energy) serves as a *natural selection* mechanism. Biological entities naturally strive to gain energy by seeking and consuming food. Similar to an entity in the biological world, each cyber-entity in the Bio-Networking Architecture may store and expend energy for living. Cyber-entities may gain energy in exchange for performing a service, and they may pay energy to use network and computing resources. The abundance or scarcity of stored energy may affect various behaviors and contributes to the natural selection process in evolution. For example, an abundance of stored energy is an indication of higher demand for the cyber-entity; thus, the cyber-entity may be designed to favor reproduction in response to higher levels of stored energy. A scarcity of stored energy (an indication of lack of demand or ineffective behaviors) may eventually cause the cyber-entity's death.

Reproduction to create diverse cyber-entity behaviors along with a mechanism for natural selection will result in the emergence of evolution to allow applications to adapt to long-term environmental change in the Bio-Networking Architecture.

3   THE BIO-NETWORKING ARCHITECTURE: ARCHITECTURAL OVERVIEW

In this and following sections, we describe details of the Bio-Networking Architecture designs. Each node in the Bio-Networking Architecture consists of the layers as shown in figure 2. A virtual machine capable of resource access control (such as the Java virtual machine) runs atop the native operating system. The Bio-Networking platform software (referred to as the *platform software* in the rest of the chapter), which provides an execution environment and supporting facilities for cyber-entities, runs using the virtual machine. The platform software also manages underlying resources such as CPU and memory. *Cyber-entities* run atop the platform software (see fig. 2). The minimum requirement for a network node to participate in the Bio-Networking Architecture is to run the platform software.

4   THE BIO-NETWORKING ARCHITECTURE: CYBER-ENTITY DESIGN

4.1   CYBER-ENTITY STRUCTURE

In the Bio-Networking Architecture design, a cyber-entity consists of three main parts: attributes, body, and behaviors (see fig. 3). *Attributes* carry information regarding the cyber-entity (e.g., cyber-entity ID, description of a service it provides, cyber-entity type, stored energy level, and age). The *body* implements a

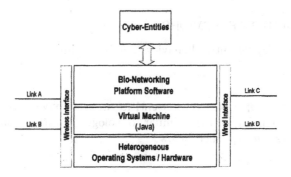

FIGURE 2   Bio-networking node.

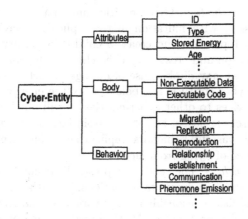

FIGURE 3   Cyber-entity components.

service that a cyber-entity provides and contains materials relevant to the service (such as data, application code, or user profiles). For instance, a cyber-entity may implement control software for a device in its body, while another cyber-entity may implement a hotel reservation service in its body. A cyber-entity that implements a web service may contain web page data in its body. Cyber-entity *behaviors* implement non-service related actions that are inherent to all cyber-entities. Examples of a behavior include migration, relationship establishment, replication, reproduction, and pheromone emission. Cyber-entity *behaviors* and *body* are described below.

4.2   CYBER-ENTITY BEHAVIORS

In the following, cyber-entity behaviors and algorithms to implement them are described.

4.2.1   Cyber-entity Behaviors.

Cyber-entities are relatively simple and autonomous and follow simple behavior rules similar to biological entities. Some of the cyber-entity behaviors are explained below.

- *Energy exchange and storage.* Cyber-entities may receive and store energy for providing services to other cyber-entities. They also expend energy. For instance, cyber-entities may pay energy units for services that they receive from other cyber-entities. In addition, when a cyber-entity uses resources on a network node, it may give energy units to the node platform software.
- *Migration.* Cyber-entities may migrate from network node (i.e., platform software) to network node (i.e., platform software).
- *Replication and reproduction.* Cyber-entities may make copies of themselves (replication), possibly with mutation in the replica's behavior. Two parent cyber-entities may reproduce and create a child cyber-entity (reproduction), possibly with mutation and crossover in the child's behavior.
- *Death.* Cyber-entities may die because of old age or lack of energy. If energy expenditure of a cyber-entity is not balanced by the energy units it receives from providing services to other cyber-entities, it will not be able to pay for the network resources it needs, i.e., it dies from lack of energy or starvation. Cyber-entities with wasteful behaviors (replicating or migrating too often) will have a higher chance of dying from lack of energy.
- *Establishing relationships.* The relationship establishment behavior enables a cyber-entity to establish and maintain relationships with other cyber-entities. Cyber-entities establish relationships under various circumstances. For example, cyber-entities collaborating to collectively provide an application are associated by an application relationship. Cyber-entities collaborating in discovery may form a discovery relationship. Cyber-entities may establish relationships with each other when they are on the same Bio-Networking platform. A relationship contains information regarding the partner cyber-entity, namely, the attributes (e.g., the cyber-entity ID, service type, age, and energy level) and the strength (a measure of usefulness) of the partner cyber-entity. Relationships are maintained autonomously by the participant cyber-entities. Such relationships may have a variety of uses, including creating applications from a group of cyber-entities and performing discovery in the Bio-Networking Architecture (to be described in section 4.4 and section 9, respectively).
- *Communication.* Cyber-entities may communicate with other cyber-entities for the purposes of, for instance, requesting a service, fulfilling a service, forwarding a discovery request, or routing packets for other cyber-entities.

$$\text{If } \sum_{i=1}^{m} W_i \cdot F_i > M, \text{ then Migrate}$$

W_1 [1011] F_i : Factors
W_2 [0001] W_i : Weights
W_3 [0110] M : Threshold

FIGURE 4   Behavior algorithm.

- *Sensing the environment.* Cyber-entities may sense their local environment. For instance, a cyber-entity may sense the local environment and learn which cyber-entities are in the environment and what services they provide. This helps a cyber-entity create a new network application or join an existing group of cyber-entities collectively providing a network application. A cyber-entity may also use sensing to obtain user information (e.g., user preference on services, and user behavior/usage patterns) and network resource information (e.g., network topology, communication link bandwidth, and CPU processing power of network nodes).

4.2.2 Cyber-entity Behavior Algorithms.   Each cyber-entity behavior may be implemented by one or more *algorithms* and a set of *factors* and *weights* which comprise each algorithm. A factor may be, for example, an operator, operand, function, or a block of code. The variables that may be changed regarding a factor are called the factor's *parameters*. For example, a weight assigned to a factor is a parameter of the factor (see fig. 4). Due to space limitations, we will only give a few examples of the cyber-entity behavior algorithms in the following.

4.2.3 Migration.   A migration behavior algorithm, for instance, may be implemented in the following manner. The migration behavior involves determining where to migrate and when to migrate, in other words, considering the cost/benefit tradeoff of migrating toward an adjacent node. The benefit of moving to the adjacent node may include lower prices for resources on the adjacent node (i.e., less energy units required to use resources on the adjacent node), proximity to an energy source or a service requester, lower delays and increased fairness for all requesting cyber-entities. Cost may include the energy units required to use the network resources for migration and higher prices for resources on the adjacent node. Therefore, possible factors in the migration behavior may include the following:

- *Energy seeking.* This factor contributes to the tendency to move toward an energy source (i.e., the cyber-entity requesting a service). (For instance, this factor may take a positive value for a network node that forwarded a service request to the node that a cyber-entity currently resides and a negative value for other nodes.)

- *Mutual repulsion.* This factor contributes to the tendency for cyber-entities of the same type to repel each other. (For instance, this factor may take a negative value for a node with a copy of a cyber-entity of the same type and a positive value for other nodes.) This factor is desirable to allow widespread distribution of cyber-entities to increase survivability and availability.

- *Resource bargain seeking.* This factor contributes to the tendency to move toward cheaper resources. (For instance, this factor may represent the resource cost of a node that a cyber-entity can migrate to.)

- *Migration cost.* This factor contributes to the tendency to move when cost of migration is small. (For instance, this factor may depend on communication resource cost such as bandwidth.)

- *Aggressiveness.* This is a factor that allows explicit adjustment of the likelihood of migration, holding all other factors constant.

One possible algorithm for a migration behavior is that if the benefit of migrating to a particular node (e.g., a weighted sum of the energy-seeking factor and the mutual repulsion factor) exceeds the migration cost (e.g., the migration cost factor) beyond a predetermined threshold value (i.e., the aggressiveness factor), then the cyber-entity migrates to that node. Other algorithms (e.g., migrating toward the source of service requests when the cost of migration is less than the cost of resources on the current node, and migrating only when the number of replica cyber-entities in the destination area is less than a certain threshold (to avoid crowding)) are also possible. These migration algorithms are only an example, and other algorithms for cyber-entity migration are possible and are implemented in a similar manner using various factors.

Please note that the list of factors provided above is not an exhaustive list of migration behavior factors. Other possible factors may include, for instance, resting factor (that contributes to the tendency for cyber-entities to stay at the current node for a certain time period so that unnecessary migration of cyber-entities and the resulting energy starvation can be prevented) and request rate factor (that contributes to the tendency for cyber-entities to stay on the current node when the arriving rate of user requests is large).

4.2.4   Replication.   Some possible factors affecting the replication behavior are:

- *Stored energy level.* A high stored energy level implies that the service provided by the cyber-entity is in high demand and that the cyber-entity has suitable behavior algorithms. A high stored energy level may encourage the cyber-entity to replicate.

- *Replication resource cost.* A high cost of resources for replication may discourage replication.
- *Population factor.* A cyber-entity may periodically measure the population density of the cyber-entity of the same type in the local area. A lower population density may encourage replication.
- *Aggressiveness.* Like the migration behavior, replication may be controlled by an aggressiveness factor to explicitly affect the likelihood of replication.

A newly replicated cyber-entity acquires energy from its parent. The newly replicated cyber-entity also derives its behavior from its parent, with or without mutation. Details of the mutation process are described in subsection 4.2.3.

4.2.5 Reproduction. The reproduction behavior produces an offspring cyber-entity from two parent cyber-entities. The reproduction behavior involves three steps: mate selection, crossover and mutation, and birth.

The *mate selection* step makes the decision to reproduce and seeks a partner with which to reproduce. The decision to reproduce may be affected by the factors of stored energy, resource cost, aggressiveness, and age. The first three factors are similar to the factors of replication. Age is considered in making the decision to reproduce since it may be more sensible to produce offspring only after the cyber-entity has survived to a certain age, thus demonstrating the viability of its behaviors.

Once the cyber-entity has made the decision to reproduce, the cyber-entity seeks a partner of equal or similar stature for reproduction by using the possible criteria described below:

- *Compatibility.* At the least, a cyber-entity must seek another cyber-entity of the same type. It is also possible that many versions of the cyber-entity software may be running. Thus, reproduction may be possible only between the same or compatible versions of cyber-entities.
- *Stored energy.* Stored energy is one of the measures of the cyber-entity's effectiveness. A cyber-entity with higher stored energy may be desirable.

Note that the proposed mate selection algorithm differs from global selection in conventional genetic algorithms [42], where the best individuals among the entire population are selected based on global knowledge of the performance of every individual in the population. The proposed mate selection algorithm is a local selection mechanism because cyber-entities use information in their immediate surroundings rather than globally measured information to perform mate selection. Local selection is more viable than global selection in an open distributed network environment since global knowledge cannot be obtained in such an environment.

The second step, *crossover and mutation*, mixes the behavioral parameters and factors of parents to produce new behaviors in offspring. This step produces diverse behavior patterns in child cyber-entities. Note that during crossover, there is no absolute method to determine which behavioral factors have been most successful. Thus, the factors and parameters that are interchanged are done on a random basis. Details of the mutation and crossover processes are described in subsection 4.2.3.

The last step of the reproduction process, *birth*, creates the new offspring cyber-entity. The new cyber-entity receives a new identifier, an age of zero, and an initial amount of stored energy from both of its parents. The new behaviors produced from crossover are injected into the new cyber-entity. The content of the new cyber-entity's body is application specific; it may be empty to serve as a new container, or it may be copied from one or both parents.

Due to space limitations, we only gave a small number of examples in this subsection. Other behaviors are implemented in the same manner.

4.2.6   Creating Diversity in the Cyber-entity Behavior.
We believe algorithmic diversity in behavior algorithms is a key requirement in evolution and in giving rise to useful emergent behavior. Therefore, we are investigating a wide range of behaviors (e.g., migration, reproduction, death, and other behaviors), a wide nature of behaviors (e.g., selfish, cooperative, and altruistic behaviors), and multiple algorithms to implement each behavior to ensure a sufficient degree of diversity. Behavioral diversity may be created in two manners; by causing mutation and crossover in cyber-entity behaviors in replication and reproduction, and by the human designer creating diverse behavior algorithms in cyber-entities.

In replication, a newly replicated cyber-entity may derive its behavior from its parent with mutation. During mutation, a new set of behaviors is derived from the parent's behaviors by changing weights in behavioral factors or removing/adding factors (see fig. 5). For instance, in the example migration algorithm that we described earlier, mutation may occur by changing the weight values of the factors (an energy seeking factor, a mutual repulsion factor, an aggressiveness factor), by removing a factor (for instance, a mutual repulsion factor), or by adding a new factor (for instance, a migration cost factor).

The reproduction behavior produces a child cyber-entity from two parent cyber-entities. In reproduction, crossover of cyber-entity behaviors is possible. *Crossover (with possible mutation)* mixes the behavioral parameters and factors of parents to produce new behaviors in offspring. Some possible methods for performing crossover include parametric crossover (exchanging parameters of factors) and factor crossover (exchanging factors and their associated parameters) [45] (see figs. 6 (a) and (b)). The crossover procedure may also include mutation, as in replication.

It is possible that evolution processes will take a long period of time to reach a certain level of desired optimality, thus the human designer may introduce algorithmic diversity so that the Bio-Networking Architecture can adapt and evolve in a shorter period of time. In addition, suitable cyber-entity behavior policies may be first developed and identified through simulations, and cyber-entities

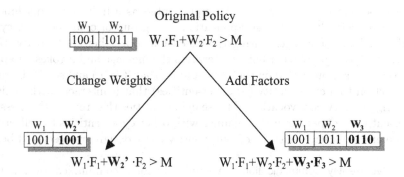

FIGURE 5   Mutation.

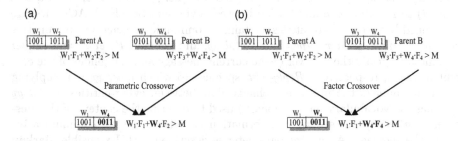

FIGURE 6   Crossover. (a) Parametric crossover. (b) Factor crossover.

with such behaviors may then be released to the Bio-Networking Architecture. This will shorten the time for cyber-entities to evolve. Note that because of the evolution mechanism inherent in the Bio-Networking Architecture, we do not need to initially develop optimal algorithms for each behavior; we expect the evolution mechanism to automatically refine and improve behavior algorithms over time.

4.3 CYBER-ENTITY BODY

In the Bio-Networking Architecture, a network application is created from a group of interacting autonomous cyber-entities [29, 30, 31, 49, 64, 65]. The network may provide multiple applications, each of which is implemented with its own group of cyber-entities. Also, a cyber-entity may belong to multiple applications. Each cyber-entity implements a functional component related to an application (i.e., service) in its body.

A cyber-entity body (service) is implemented as a finite state machine. In order to collectively provide an application as a group of cyber-entities, cyber-entities exchange messages during the execution of a cyber-entity body. Upon receiving a message, a cyber-entity interprets the message and invokes an appropriate service action (based on the current state that it is in) and sends the result of the action to a cyber-entity (or cyber-entities) that it interacts with. This, in turn, triggers service invocation of those cyber-entities that receive the message. Cyber-entities cooperate in this manner with other cyber-entities to collectively provide an application. Details of cyber-entity body design are described below.

4.3.1 Cyber-entity Communication.

A candidate for a communication language for cyber-entities to use is the Speech Act-based Agent Communication Language (ACL) [1, 16, 56] with extensions specific to the Bio-Networking Architecture [26, 32, 33]. In the ACL, we define a small number of communicative acts (such as *request, agree, refuse, inform, failure, query-if, advertise, recruit,* and *reward*) to facilitate communication between cyber-entities. Each ACL message exchanged between cyber-entities contains a communicative act and parameters such as *receiver, sender, reply-with, ontology,* and *content. Receiver* and *sender* parameters specify the receiver of the current message and the sender of the current message, respectively. *Reply-with* specifies to which message it is replying and is to manage the message exchange flow between cyber-entities. *Ontology* specifies the vocabulary set (dictionary) used to describe the content of the message. *Content* specifies data or information associated with a communicative act in the message. A *content* parameter is described with Extensible Markup Language (XML) [80]. The semantics of a *content* parameter are specific to each cyber-entity because different services have different semantics. For instance, the attribute of "date" may be interpreted as a birthday in a greeting card delivery service, while it may be interpreted as a part of schedule in a Personal Information Management (PIM) service. Upon receiving an ACL message, a cyber-entity interprets the communicative act and parameters contained in the message, and invokes an appropriate service action.

An example sequence of communicative acts is described below. A cyber-entity sends *request* to cyber-entities that it has a relationship with and asks them to invoke their service actions. Each receiver cyber-entity responds with either *agree* or *refuse* (to invoke service) to the sender cyber-entity. If the receiver cyber-entity is willing to accept *request* from the sender, it responds with *agree*, invokes its service action and then sends *inform* to notify the sender cyber-entity of the result of the service action. (If the service action fails, it sends *failure* to the sender cyber-entity to inform that the action failed.)

Other examples of how cyber-entities use communicative acts include the following. A cyber-entity may send *advertise* to notify nearby cyber-entities of its existence. A cyber-entity may send *recruit* to solicit other cyber-entities to respond to its message. (Please see section 4.4. for an illustration of how *advertise* and *recruit* are used to establish a relationship between cyber-entities.) When a user receives a service, the user returns *reward* for the service. *Reward* back propagates to all cyber-entities that participated in the service interaction.

4.3.2 Cyber-entity Body Design.

The Bio-Networking interaction protocol (IP) defines a sequence of interactions (invocation of service actions and exchange of the Bio-Networking ACL messages) among cyber-entities to provide an application [26, 32, 33]. In the Bio-Networking Architecture, a cyber-entity implements its service in the body. A cyber-entity body defines the IP for the service it provides and is implemented as a finite state machine that consists of states and state transition rules. A state implements an atomic action (i.e., an atomic service) and message exchanges associated with the action (to allow inputting data to and outputting data from a given action in a given state). For instance, a cyber-entity implementing a hotel reservation service may contain a state consisting of the action "check room availability" and associated message exchanges such as "date" and "the number of customers" as input data, and produce "hotel name," "assigned room number," and "room rate" as output data. A state transition rule associated with a state specifies the next state to transit to. Different state transition rules may be defined for different outcomes of a given action. When an action in a given state is completed, the current state moves to the next state based on the state transition rule.

Upon receiving a message from another cyber-entity, a cyber-entity examines parameters of the incoming message and invokes an appropriate service action. For instance, if the parameter *:in-reply-to* is set in the incoming message, it is in response to a previously transmitted message, and the cyber-entity invokes the corresponding service action associated with the state where the previous message was transmitted. If the parameter *:in-reply-to* of the incoming message is null, the incoming message is the first message from the sender cyber-entity, and the receiver cyber-entity examines the initial state of each and every IP that it implements. The service action of the initial state of the IP that can take the incoming message as its input is then invoked.

4.4   CREATION AND CUSTOMIZATION OF APPLICATIONS

In the Bio-Networking Architecture, cyber-entities interact and collectively provide an application. In providing applications, a cyber-entity in the Bio-Networking Architecture first establishes an application relationship through interaction according to its own application relationship establishment policy. A cyber-entity then selects a cyber-entity (or cyber-entities) to interact with from those that it has an application relationship with. As will be described later in this subsection, the application relationship contains service-related attributes of the relationship partner cyber-entities (such as the cyber-entity ID, service IP name, and service properties) and the strength of the application relationship to the partner cyber-entities.

The strength of an application relationship is modified based on the satisfaction degree of a user who received the application in a feedback loop. When the application relationship strength among cyber-entities that collectively provide an application becomes greater than a predetermined threshold value, they form an application group so that the application may be retrieved easily at a later time.

In the following, we describe the process in which an application is created from a group of interacting cyber-entities (i.e., what information is kept in an application relationship table at a cyber-entity, how cyber-entities may establish application relationships, how an interaction partner cyber-entity may be selected, how cyber-entities interact to collectively provide an application, how an application relationship strength is adjusted based on user feedback, and how an application group is created).

4.4.1 Application Relationship. In the Bio-Networking Architecture, a cyber- entity establishes application relationships with other cyber-entities and records useful interactions in the form of application relationship. An application relationship may be viewed as a cyber-entity's information cache regarding interactions with other cyber-entities.

Table 1 shows examples of application relationship attributes stored in an application relationship record at a cyber-entity. *CE-id* is used to uniquely identify a relationship partner cyber-entity (CE). *IP-name* specifies an interaction protocol of self (i.e., its own interaction protocol) to use to interact with a relationship partner cyber-entity. *Service-properties* stores information regarding the service a relationship partner cyber-entity provides (such as semantic information about the service). For instance, *service-properties* may store the type and the keywords of the service that a relationship partner cyber-entity provides or the keywords stored in a query issued by the relationship partner cyber-entity. *Access-count* counts the number of messages exchanged with a relationship partner cyber-entity. *Strength* indicates the usefulness of a partner cyber-entity and is used to select an interaction partner cyber-entity.

4.4.2 Application Relationship Establishment. Cyber-entities establish application relationships through interaction. For instance, a cyber-entity that has just migrated to a new node may broadcast *advertise* (a Bio-Networking ACL message) to establish application relationships with nearby cyber-entities. When a nearby cyber-entity receives *advertise*, it creates a new entry in the application relationship record and stores the CE-id of the sender cyber-entity. Additional information obtained through interaction regarding a relationship partner cyber-entity may be stored in *service-properties* in the relationship record. In order to find and establish an application relationship with a specific partner (e.g., a cyber-entity of a certain service type or with certain attributes), a cyber-entity may broadcast *recruit*, specifying conditions, and attributes of a desired relationship partner (e.g., service type and/or attributes required for a partner cyber-entity).

TABLE 1   Example attributes of an application relationship record at a cyber-entity.

Attribute Names	Meaning
CE-id	A globally unique identifier of a relationship partner cyber-entity
IP-name	An interaction protocol (IP) of self to use to interact with a relationship partner cyber-entity
Service-properties	Information regarding the service of a relationship partner cyber-entity
Access-count	The number of interactions with a relationship partner cyber-entity
Strength	Indication of the usefulness of a relationship partner cyber-entity

A cyber-entity that receives *recruit* and matches the specified conditions, responds with *inform*, containing its own attributes. Upon receiving *inform*, the sender cyber-entity of *recruit* establishes a relationship with the cyber-entity that responded with *inform* and stores necessary information regarding the partner cyber-entity in the application relationship record.

4.4.3   Interaction Partner Selection.

In deciding which cyber-entity to interact with, a cyber-entity examines its application relationship partner cyber-entities and chooses the cyber-entities that fit its criteria. For instance, a cyber-entity may specify one or more relationship attributes as keys and select cyber-entities whose relationships match the specified keys. As an example, assume that *service-properties* in the application relationship record of a cyber-entity includes "service type" of relationship partner cyber-entities. This cyber-entity may select as an interaction partner the cyber-entity of the service type "travel." If there are multiple cyber-entities that match the keys, they can be further narrowed based on the strength of the application relationship. For instance, cyber-entities may be selected proportionally to the strength of their application relationships so that cyber-entities with stronger relationships are more likely to be selected as interaction partners.

By selecting interaction partner cyber-entities based on the application relationship strength, cyber-entities self-organize to provide desirable and customized applications. Cyber-entities with useful services (i.e., cyber-entities with strong application relationships) are likely to be selected as interaction partners, and those with weak relationships are not likely to be selected.

4.4.4   Service Interaction among Cyber-entities.

Figure 7 shows how cyber-entities interact and collectively provide an application using application relationships.

Actions and state transition rules associated with each state in an IP of a cyber-entity service (body) are implemented by the cyber-entity designer and registered with a state model in the cyber-entity service (body) (depicted as "(1) register" in fig. 7).

Platform software's *communication service* receives service-related Bio-Networking ACL messages from other cyber-entities and dispatches them to the cyber-entity service (body) (depicted as "(2) dispatch" in fig. 7). (Designs of the platform software will be described in detail in section 5.)

Upon receiving a Bio-Networking ACL message, the cyber-entity invokes an *IP engine* (in cyber-entity service (body)) that manages the corresponding session[2] in the following manner. The *IP engine* controls the execution of a state model and is created for each cyber-entity interaction session. If the value of the parameter *:in-reply-to* in an incoming message is non-null (i.e., if the two cyber-entities have interacted before), the state of the session where the previous message was sent is invoked. The *IP engine* then invokes the action associated with the current state (depicted as "(6) act" in fig. 7). If the value of the parameter *:in-reply-to* is null (i.e., if the incoming message is the first message sent from the sender cyber-entity), a new *state model* is selected by examining the parameters of an incoming message, and a new session is invoked for the selected *state model* (depicted as "(3) get" and "(4) invoke" in fig. 7). The *IP engine* then invokes the action associated with the current state.

The *IP engine* also creates an application relationship record regarding the cyber-entity that sent the Bio-Networking ACL message (depicted as "(5) set" in fig. 7). The *service-properties* attribute of the application relationship record stores semantic information about the service that the application relationship partner cyber-entity provides (depicted as "(7) set" in fig. 7). Other attributes such as *CE-id* and *Access-count* are also stored in the application relationship record. This application relationship record may be examined by an action that is associated with a current state to select a receiver cyber-entity for an outgoing message (depicted as "(8) select" in fig. 7). The outgoing message may be either unicast, multicast (depicted as "(9) convey" in fig. 7) or broadcast (depicted as "(10) spread" in fig. 7) to the receiver cyber-entities.

4.4.5 Application Relationship Strength Adjustment.

In the Bio-Networking Architecture, the strength of an application relationship is adjusted based on the degree of satisfaction indicated by the user who received the application. In the Bio-Networking Architecture, a user creates a *reward* message and indicates his/her *happiness* (i.e., the degree of his/her satisfaction with the application he/she received) in a *reward* message. A *reward* message back propagates the path that cyber-entities used to interact and provide an application. In order to allow for back propagation of a *reward* message, each cyber-entity involved in providing an application records the previous and next cyber-entities in the interaction sequence in providing the application. Upon receiving a *reward* mes-

---

[2]A session refers to a series of Bio-Networking ACL message exchanges between two interacting cyber-entities.

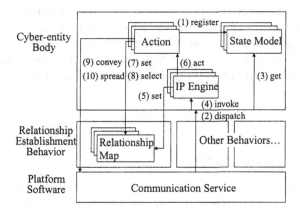

FIGURE 7   Interactions to provide an application.

sage, each cyber-entity on the back propagation path adjusts the strength of application relationships based on the *happiness* value in the *reward* message. If a user is satisfied with the application, relationship strength is increased; if a user is not satisfied with the application, relationship strength is decreased. If a user is neutral, there is no change in relationship strength. Cyber-entities that collectively provide a popular application (i.e., an application that a number of users like) will receive a number of positive happiness feedback from users, and their relationship strengths will increase, while relationships among cyber-entities that provide unpopular applications are weakened.

4.4.6   Application Group Formation.   When relationship strengths between cyber-entities collectively providing an application exceed a predetermined threshold, such cyber-entities may form a permanent group to ensure repeatability of the application. When a user feedback propagates through cyber-entities involved in an application, it can examine the relationship strength among cyber-entities. Once a group is formed, a unique group ID is assigned to the group, and each cyber-entity in the group stores the group ID, a previous cyber-entity (in interactions), and a next cyber-entity. With such entries in all group member cyber-entities, it is possible to locate all members of the group and to traverse the application along specific cyber-entities in a particular order.

4.5   EXAMPLE NETWORK APPLICATIONS IN THE BIO-NETWORKING ARCHITECTURE

As discussed earlier in sections 2 and 4.4, in the Bio-Networking Architecture, a network application is provided by collaborating cyber-entities. Each cyber-entity implements functionality related to the network application and follows a set of simple behavior rules. In the Bio-Networking Architecture, desirable behaviors and characteristics emerge in the application from the collective actions and interactions of its cyber-entities. Some examples of network applications that the Bio-Networking Architecture can enable are explained below [25, 26, 32, 33, 64].

4.6   CUSTOMIZED SERVICES

In this example, a network application that the Bio-Networking Architecture provides adjusts to the preference and usage patterns of an individual user. As an example, consider the following interaction sequence in a house. Kentaro, a son of the house, returns home. A cyber-entity providing sensory information (a sensor cyber-entity) detects the movement of an entrance door of the house and broadcasts a "door is open" signal. A cyber-entity that controls a video camera (a video camera cyber-entity) receives the signal and starts videotaping a moving object, sending the image to an image analyzer cyber-entity running on a PC on Shotaro's (Kentaro's brother's) desk in the house. The image analyzer cyber-entity analyzes the incoming video signal and detects that the person entering the door is a son of the house, Kentaro. The image analyzer cyber-entity then broadcasts a "Kentaro is back at home" signal. A cyber-entity that controls a stereo set in the living room receives the signal, finds a nearby cyber-entity that stores Miles Davis music MP3 software, and starts playing jazz music. Kentaro, however, does not like jazz and indicates so by turning off the stereo or by turning down the stereo volume. The relationship strength between the stereo set and Miles Davis MP3 software cyber-entities is weakened. The stereo set cyber-entity invokes a discovery mechanism, finds alternate music (e.g., a cyber-entity that stores Rita Coolidge music MP3 software), and starts playing the music. Kentaro this time turns the volume up, indicating that he likes the music. The relationship between the stereo set cyber-entity and the Rita Coolidge MP3 software cyber-entity is now strengthened. Next day, when Kentaro returns home from his school, the same sequence of cyber-entity interactions occurs, and the stereo set in the living room again first plays Miles Davis jazz music. Kentaro, again, indicates he prefers other types of music by turning off the stereo set. This further weakens the relationship between the stereo set and Miles Davis MP3 software cyber-entities. The stereo set again invokes a discovery mechanism, finds the Rita Coolidge MP3 software cyber-entity and plays the music. Kentaro, again, indicates he likes the music by turning up the volume (or by simply not turning off the stereo set), strengthening the relationship between the stereo set and Rita Coolidge MP3 software cyber-entities. After a few days of the same sequence of cyber-entity interactions, the stereo set cyber-entity automatically starts playing the Rita Coolidge music when Kentaro enters the house.

4.7   CREATION OF NEW SERVICES

In the example in section 4.5.1, a network application is customized based on the preference and usage patterns of an individual user. In the example below, a new network application emerges in the Bio-Networking Architecture based on the preference and usage patterns of a group of users.

As an example, consider a popular public spot such as the New York City's Times Square (and the building with the large TV screen in Times Square) and a busy street in front of a large TV screen on the Alta Building in Shinjuku, Japan. A number of people (users) visit Times Square (or the busy street in front of the Alta TV screen), stay there for awhile (doing, for instance, window shopping, chatting with friends, having some coffee at a cafe), and then leave. Assume that many who visit Times Square carry mobile phones or PDAs that are capable of communicating with other mobile phones and PDAs in an ad-hoc manner. Assume also that each of these mobile phones and PDAs runs a cyber-entity (a user cyber-entity), monitoring and storing the user's profile (e.g., user's preference and usage patterns of applications). Assume that the large TV screen in Times Square runs a cyber-entity (a TV screen cyber-entity) that communicates with nearby user cyber-entities. Assume also that each shop in the Times Square area implements a cyber-entity (a video commercial cyber-entity) that stores a video commercial clip of the shop and its merchandise and runs this cyber-entity on a computer in the shop. In addition, some users may have downloaded a video commercial cyber-entity in their mobile phones and PDAs from where they visited earlier in the day.

Let's consider the following cyber-entity interaction sequence. The TV screen cyber-entity in Times Square communicates with user cyber-entities within its communication range and collects user preferences stored in the user cyber-entities. Based on the collected user preference information, the TV screen cyber-entity searches for video commercial cyber-entities that advertise shops and merchandise matching the preferences of users in Times Square and starts displaying such video commercial clips.

Assume that, although people constantly come and go in Times Square, the types of shops and merchandise that the crowd in Times Square prefers do not change significantly for a certain time period. (For instance, there is a large number of tourists visiting Times Square on a daily basis, and they tend to seek information about tourist gifts. There are many after-theater people from the nearby Broadway theater district who enter and leave Times Square, looking for restaurants for an after-theater dinner in the evening.) Reflecting such relatively stable changes in the user preference of the crowd in Times Square, some video commercial clips are played on the TV screen more often than others, and through this frequent interaction with the TV screen cyber-entity, they start establishing and strengthening the relationships among themselves. Once the relationship strength becomes sufficiently strong, displaying one video commercial clip will result in displaying other video commercial clips that match the user preference of the crowd in Times Square.

4.8 APPLICATION DESIGN AND IMPLEMENTATION

We are currently designing an application of automated ticket sales using the Bio-Networking Architecture. In this application, a group of cyber-entities interact and dynamically create relationships to provide a ticket sales service. As relationships are dynamically created, a web of cyber-entities emerges, and ticket sales services dynamically adjust to user preferences. Please refer to [25, 26, 32, 33, 64] for detailed descriptions of the services that we are currently implementing.

5 THE BIO-NETWORKING ARCHITECTURE: PLATFORM SOFTWARE DESIGN

The Bio-Network platform software (hereafter simply referred to as the platform software) enables a network node to participate as a node in a virtual network of Bio-Networking enabled nodes [67]. This virtual network may be overlaid on top of the conventional Internet protocol network, similar to the way in which the virtual multicast backbone, or mbone, operates over the current unicast Internet protocol network. The platform software also provides an execution environment for cyber-entities. Some of the facilities implemented by the platform software, such as execution environment, migration, and lifecycle management, are similar to those of existing mobile agent platforms [10]. However, the platform software also implements additional facilities not found in existing mobile agent platforms, such as fine-grained resource control, cyber-entity energy management, and topology management (of the virtual network of Bio-Networking enabled nodes).

The platform software may also implement some of the cyber-entity behaviors that we described in section 4.2 (as opposed to implementing them in a cyber-entity) due to concerns regarding trust and efficiency. The platform software, like the operating system kernel, can be trusted and provides optimized versions of commonly used functions. For instance, the platform software may implement the energy unit variable for all cyber-entities on that node [61, 62]. This prevents malicious cyber-entities from cheating and arbitrarily modifying their own energy unit variables. The platform software may also maintain a relationship structure that connects cyber-entities collectively providing a network application; functionality that all cyber-entities require. When the platform software implements this commonly used functionality, it frees all cyber-entities from having to do so, therefore reducing the cyber-entity size and complexity.

The disadvantage of implementing functionality in the platform software is that the functionality cannot adapt or evolve through the inherent mechanisms of the Bio-Networking Architecture. We are currently exploring and evaluating both

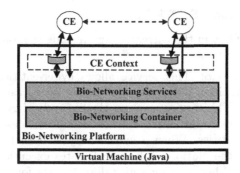

FIGURE 8   Platform design.

alternatives (implementing in platform software versus in cyber-entity behavior) for a variety of functionalities.

In the current design [66], the platform software consists of cyber-entity (CE) context, Bio-Networking services and a Bio-Networking container and runs using the virtual machine (see fig. 8). Cyber-entity context provides cyber-entities with references to services available from Bio-Networking services. Bio-Networking services provide a set of general-purpose runtime services that are frequently used by cyber-entities. These services abstract low level operations such as cyber-entity life cycle management, resource allocation to cyber-entities, and migration of cyber-entities. Bio-Networking services relieve cyber-entities of low-level operations and also allow cyber-entities to be lightweight by separating them from routine work. A Bio-Networking container provides the functionality that is required to provide Bio-Networking services. The Bio-Networking container functionality provided in the current design includes such functions as registering a newly created cyber-entity in a local registration table, maintaining a cyber-entity reference ID, and communication primitives for cyber-entities. In the following, we describe facilities that are likely to be implemented in the platform software.

5.1   CYBER-ENTITY CONTEXT

Cyber-entity context is an entry point for a cyber-entity to access Bio-Networking services. It determines if a Bio-Networking service requested by a cyber-entity is available, and if it is, it obtains a reference to the service. CE context is created and associated with each cyber-entity implicitly (automatically) by the Bio-Networking life cycle service, when a cyber-entity is created, replicated, reproduced or migrated from another node.

5.2   BIO-NETWORKING SERVICES

Bio-Networking services in the current design include the following services:

- *Communication service.* This service allows cyber-entities to communicate with each other in either a unicast, multicast or broadcast manner.
- *Life cycle service.* This service manages the life cycle of cyber-entities. It provides the operations to initialize, activate, deactivate, destroy, replicate, and reproduce cyber-entities. When a cyber-entity is born (through replication or reproduction), the platform software allocates execution resources to the new cyber-entity. When a cyber-entity dies, the platform software frees the execution resources used by that cyber-entity.
- *Directory service.* This service runs on per-node basis and keeps the information regarding the cyber-entities that exist on a local node.
- *Migration service.* This service is responsible for sending and receiving a cyber-entity to and from another node. The migration facility in the platform software is similar to those found in existing mobile agent platforms [27, 10, 28]. The platform software on the source node contacts the platform software on the destination node to establish a reliable connection. The destination node then reserves execution resources such as a thread and memory space. After the cyber-entity's code and execution state are transported to the destination node, execution resources on the source node are freed.
- *Energy management service.* This service allows a cyber-entity to pay energy for a service provided by another cyber-entity and for the resources that the cyber-entity uses. It manages the energy level of cyber-entities on the platform. This service also manages the energy level of a cyber-entity when it migrates, replicates and reproduces.
- *Relationship management service.* This service allows cyber-entities to establish, examine, update, and destroy relationships with one or more cyber-entities.
- *Discovery service.* This service allows a cyber-entity to discover another cyber-entity on a remote platform. Discovery in the Bio-Networking Architecture is through relationships between cyber-entities.
- *Resource sensing service.* This service senses the type, amount, and cost of resources available on both a local platform and neighboring platforms. Types of resource that may be sensed include physical resources (i.e., CPU time and memory space) and logical resources (i.e., thread and transport connection).
- *Resource allocation service.* This service assigns physical and logical resources to a cyber-entity. This service controls the cyber-entity's access to system resources such as CPU usage (via scheduling), memory usage (via controlled memory allocation), network communication, and file system access. The amount and type of access granted are determined by the cyber-entity's privileges. Privileges may be defined by the type of the cyber-entity or by the amount of energy the cyber-entity has paid.

5.3 BIO-NETWORKING CONTAINER

Bio-Networking services run on a Bio-Networking container. A Bio-Networking container provides low level operations to maintain the platform software, and it provides the following functionalities:

- *Cyber-entity registration and unregistration.* This functionality maintains a cyber-entity registration table. When a cyber-entity is created, a Bio-Networking container registers the cyber-entity in its registration table so that it can route incoming requests to the cyber-entity. When a cyber-entity is destroyed, a Bio-Networking container unregisters the cyber-entity from its registration table.
- *Cyber-entity reference management.* A Bio-Networking container creates, modifies, and destroys cyber-entity references. A cyber-entity reference is a pointer to a cyber-entity to allow other cyber-entities to locate and communicate with the cyber-entity. It is created when a cyber-entity is created and registered in a Bio-Networking container. When a cyber-entity migrates to another node, the reference of the cyber-entity at the original node is destroyed, and the Bio-Networking container on the destination node assigns a new reference to the cyber-entity.
- *Parsing of requests and events.* A Bio-Networking container receives and parses incoming requests and events, and then routes them to target cyber-entities by referring to a cyber-entity registration table.
- *Cyber-entity activation and deactivation.* A Bio-Networking container is responsible for activation and deactivation of cyber-entities. In activating a cyber-entity, a Bio-Networking container loads a cyber-entity from a persistent storage space (e.g. file or database) to memory. In deactivating a cyber-entity, a Bio-Networking container unloads a cyber-entity from memory to a persistent storage space.
- *Resource management.* A Bio-Networking container is aware of the types, amounts, and cost of available resources. It accounts for and allocates physical and logical resources to cyber-entities.

We have completed the design of the platform software and are currently implementing the platform software with the Java programming language.

6 OTHER BIOLOGICAL CONCEPTS AND MECHANISMS USED IN THE BIO-NETWORKING ARCHITECTURE

In addition to those explained in section 2, we consider borrowing a spectrum of additional concepts and mechanisms from the biological systems. Some of them are explained below. Due to space limitations, only a small number of examples are listed below.

- *Pheromone emission.* Cyber-entities may emit pheromones to their neighboring nodes to indicate their presence to other cyber-entities. Pheromones contain information regarding the emitter's identity and its attributes. Pheromones are emitted with a certain strength and propagation radius and may dissipate with time. This mechanism may have a variety of uses, including improving the performance of discovery.
- *Bacterial conjugation.* Conjugation provides a mechanism by which bacteria exchange genetic information while alive. This allows the rapid spread of beneficial genetics throughout a population. Cyber-entities can employ similar mechanisms to increase the rate of evolution.
- *Cell differentiation and developmental biology.* Cells within organisms often differentiate to provide specialized functionality, allowing the cell to have improved performance for a particular task. The different types of cells collectively provide functionality that no single type of cell can provide. Developmental biology [79] describes mechanisms that coordinate differentiation based on local information. Similarly, cyber-entities can be designed to sense nearby cyber-entities or nearby platforms and to differentiate into specialized types of behaviors for improved performance (e.g., to meet a particular pattern of user demand or pattern of platform resources).
- *Symbiosis and altruism behaviors.* Cyber-entities may be designed such that one cyber-entity benefits from another (commensalism) or such that two cyber-entities benefit from each other (mutualism). For instance, as a result of differentiation described above, different cyber-entities may develop specialization in finding different types of cyber-entities. Such specialized entities may collaborate by delegating appropriate search packets among themselves. By doing so, cyber-entities may provide better discovery service or acquire more energy than if each cyber-entity supported all searches.
- *Seasonal patterns.* Animals may perform different actions in response to seasonal resource patterns. Certain animals may migrate toward likely resources without explicit indication that such resources exist. Similarly, cyber-entities may be pre-programmed or use history to perform long-range migration behavior to place them in areas that are likely to have higher user demand. Other animals may hibernate when resources become scarce. Cyber-entities could also hibernate by selecting energy-conservative behavior policies or by moving into more energy efficient platforms. This would keep entities available, but more resource efficient when user demand is low.

7 EMERGENT BEHAVIORS AND CHARACTERISTICS IN THE BIO-NETWORKING ARCHITECTURE

We expect useful emergent behaviors and characteristics to emerge from simple behaviors of cyber-entities as described in section 2. Some such emergent behaviors are described below.

- *Adaptation* of cyber-entities to heterogeneous and dynamic network conditions emerges. Cyber-entities can be designed to migrate toward the source of service requests (i.e., a user) when the resource cost at the migration destination node is less than a given threshold, while avoiding areas of the network where resource costs are high. Cyber-entities can be, at the same time, designed to replicate when demand is high (i.e., when their stored energy levels are high), and to die when demand is low (i.e., when their stored energy levels reach zero). The emergent result of these simple individual behaviors is that cyber-entities adapt their population to the number of service requests, the source of service requests, and the cost of network resources. (See section 8 for the adaptation capability of cyber-entities.) The *adaptation* of applications (provided by collaborating cyber-entities) to diverse user preferences emerges through the relationship establishment behavior of cyber-entities in the group (as explained in section 4.4).
- *Evolution* of cyber-entities emerges from the continuous development of new behavioral algorithms formed through mutation and crossover. Death from old age and lack of energy (resulting from unfit cyber-entity behaviors and an unpopular service implemented in a cyber-entity body) steers the cyber-entity population toward more effective behaviors. It is also expected that diversity in cyber-entity behavior coupled with variations in conditions in different parts of the network will result in localization, which causes some cyber-entities to evolve behaviors suitable for their local environmental conditions, while other cyber-entities will evolve behaviors more suitable for their own local environment. For example, the natural selection mechanisms inherent in the Bio-Networking Architecture may cause cyber-entities in a resource-abundant part of the network to evolve toward a higher rate of reproduction, while in a resource-limited part of the network, cyber-entities evolve toward a lower rate.
- *Scalability* is achieved because cyber-entities act autonomously and on a local basis using only local information. As the number of cyber-entities grows, performance bottlenecks do not appear around a critical or controller cyber-entity.
- *Security and survivability* in the Bio-Networking Architecture are provided by a variety of mechanisms. Such mechanisms include replication of cyber-entities and algorithmic diversity in cyber-entity behavior. Replication increases the survivability of cyber-entities from attacks or failures of the platform software. Algorithmic diversity (i.e., implementing a behavior with different algorithms, or with the same algorithm but with different parameters) can be introduced by the designer or automatically generated through the mutation and crossover mechanisms in the Bio-Networking Architecture. Since attacks and failures sometimes depend on an exact algorithmic sequence or parameter, some cyber-entities may be unaffected by a particular attack. In addition, cyber-entities can be designed to spread widely to improve chances of survival from large-scale network attacks or failures.

- *Self-organization* occurs in the Bio-Networking Architecture as the result of the relationships formed by cyber-entities. Cyber-entities may acquire relationships and modify their strengths based on simple policies, resulting in emergent high-level organization. For instance, in discovery, cyber-entities self-organize, and certain properties of the relationship network may emerge. Cyber-entities may have differing abilities to perform searches as a result of different environments where they exist and different relationship qualities that they have. Search strength may be modified based on history, and relationships to cyber-entities that are poor at performing search may be replaced by relationships to cyber-entities that have performed well at searches. This will result in a search relationship topology clustered around useful cyber-entities.

8 PRELIMINARY SIMULATION STUDY OF THE BIO-NETWORKING ARCHITECTURE

8.1 APHID, A WEB CONTENT DISTRIBUTION APPLICATION BASED ON THE BIO-NETWORKING ARCHITECTURE

We are currently simulating a Bio-Networking Architecture application called Aphid to demonstrate basic features of the Bio-Networking Architecture and how it may be used to construct a network application [72, 73, 74, 75, 76]. Aphid is a scalable, adaptive, and survivable/available web content distribution application. The Aphid application consists of multiple Aphid cyber-entities. Aphid cyber-entities have behaviors described in section 4.2. Aphid cyber-entities accept requests for web pages and deliver them using the HTTP protocol. The body of an Aphid cyber-entity may contain all the web pages or just the most popular pages of a web site. Simulation data obtained to date have shown that the Aphid exhibits unique and desirable features such as scalability, adaptability, and survivability/availability. In the following, we will show some of the simulation results.

8.2 SIMULATION PARAMETERS

The simulation results presented in the following are based on the 50-node network topology shown in figure 9. The topology represents a very simplified version of the global Internet. Each simulation run simulates 24 hours of real time. The number of users during the simulation runs changes and is shown in figure 10. When simulations start, users are placed at randomly selected nodes in the network. Users request one web page per second. Each user's requests go to the Aphid cyber-entity closest to the user. When the request is made, users pay a fixed amount of energy units (40) to the cyber-entity.

In the simulation results we show below, the bodies of all Aphid cyber-entities contain copies of the same web page. Each Aphid cyber-entity pays 700 energy units per second to the Bio-Networking platform software it is on. The

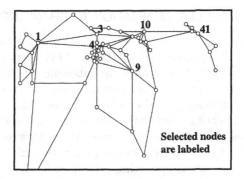

FIGURE 9 Network topology.

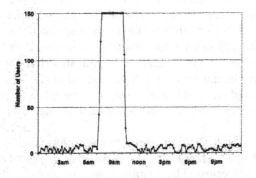

FIGURE 10 User demand during simulation.

700 energy units buy enough resources (CPU, memory, and network bandwidth) for the cyber-entity to service 20 requests per second. If a cyber-entity replicates (i.e., if a cyber-entity reproduces asexually), it must pay an additional 525 energy units. If it migrates to an adjacent node (i.e., platform software), it must pay additional 11,200 energy units. In replication (i.e., asexual reproduction), a parent cyber-entity gives a minimum of 2.5 million energy units to its child.

In order to simplify the simulations, cyber-entities replicate (i.e., reproduce asexually) without mutation of the behaviors. Sexual reproduction, crossover, and mutation are not simulated. As a result of these simplifications, evolution does not occur. Although the scenarios used in the simulations are specific to Aphid and have rather restrictive assumptions (e.g., no sexual reproduction and no service provided collectively by a group of cyber-entities are simulated), we believe that preliminary simulation results we have obtained provide a general understanding of the Bio-Networking Architecture.

8.3   SIMULATION RESULTS; ADAPTATION TO CHANGING USER DEMAND

To demonstrate the Aphid cyber-entity's ability to adapt their population to changing user demand, we simulated two types of cyber-entities, CE1 and CE-Best. Both types of cyber-entities have the following set of behaviors: energy exchange and storage, replication (i.e., asexual reproduction), death, and communication. Notably missing from the cyber-entity behavior is migration.

In the CE1 cyber-entity's replication behavior (i.e., asexual reproduction), the *reproductionRequestRate* factor encourages the replication (i.e., asexual reproduction) behavior if the cyber-entity's request rate is high and/or if the rate of change in the request rate is increasing. The *reproductionStoredEnergy* factor strongly inhibits the replication (i.e., asexual reproduction) behavior when the cyber-entity's energy level is below the minimum amount that it must give to its child. In the CE1 cyber-entity's death behavior, the *deathRequestRate* factor encourages the death behavior if the cyber-entity's request rate is low. The *deathAge* factor inhibits the death behavior if the cyber-entity was recently born. The *deathPopulation1* factor inhibits the death behavior if the cyber-entity has no relationships with any other cyber-entities. (When a cyber-entity has no relationships with any other cyber-entities, it indicates that this cyber-entity is the only cyber-entity in the network.) Note that all of the factors used by CE1 cyber-entities are based on local information, e.g., the cyber-entity's request rate, the cyber-entity's energy level, the number of relationships that the cyber-entity has.

CE-Best cyber-entities provide the best case performance of any adaptation algorithm. CE-Best cyber-entities do not follow the biological principles of the Bio-Networking Architecture. The simulator uses its global information on the number and location of users to centrally control the number and location of CE-Best cyber-entities.

Figure 11 shows the cyber-entity population during the simulation runs. In figure 11, both CE1 and CE-Best cyber-entities reproduce in order to adapt to the increasing user demand at 7 a.m. Both adapt to the decreasing user demand at 10 a.m. The population curve for the CE-Best cyber-entities represents the optimal population curve because the simulator can determine exactly how many best cyber-entities are needed throughout the simulation period. The shape of the population curves in figure 11 shows that the CE1 cyber-entities are able to adapt to increasing and decreasing user demand.

8.4   SIMULATION RESULTS; ADAPTATION TO THE LOCATION OF USERS IN THE NETWORK

It is desirable for the Aphid cyber-entities to adapt to the location of the users by migrating toward them. The result of cyber-entity migration is a reduction in the average number of network hops between users and their closest cyber-entities. To demonstrate the Aphid cyber-entity's ability to adapt to the location of users in the network, we have added migration behavior to CE1 and created a new

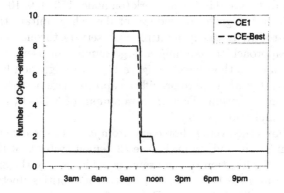

FIGURE 11   Cyber-entity population.

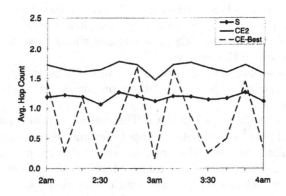

FIGURE 12   Hop count comparison.

type of cyber-entity, CE2. CE2 cyber-entity's migration behavior contains the following factors. The *migrationRequestPercentage* factor encourages migration to a neighboring node if more than a certain percentage of the cyber-entity's total requests is coming from the direction of the node. The *migrationRest* factor inhibits the migration behavior if the cyber-entity has migrated recently. The *migrationRepulsion* factor encourages the cyber-entity to migrate away from its current platform software, if there is another Aphid cyber-entity on the same platform. Consistent with biological principles, these factors are based only on local information.

As a comparison base, the Static6 cyber-entities and CE-Best cyber-entities are also simulated. The Static6 cyber-entity is actually 6 static cyber-entities that are strategically distributed in the network (on nodes 1, 3, 4, 9, 10, and 41). Static cyber-entities do not migrate, reproduce, or die. They imitate the web servers that are in use today. Locating multiple web servers throughout the network is one possible approach to handling a large number of users. CE-Best cyber-entities are the same as those used in figure 11, with migration behavior added. CE-Best cyber-entities are used to provide an approximate best-case performance of any adaptation algorithm. There is a minimum of 3 CE-Best cyber-entities throughout the simulation period.

Figure 12 shows the average hop count from users to the cyber-entity closest to them between 2 a.m. and 4 a.m. Figure 12 demonstrates that the CE2 cyber-entities adapt to the location of the users, as evidenced by the smaller average hop counts. The average hop count for best cyber-entities fluctuates between 0.1 and 1.7 for the following reason. Between 2 a.m. and 4 a.m., the number of users varies between 0 and 10. When there are 0 to 3 users in the network, the 3 CE-Best cyber-entities are placed on the same node as the users, resulting in an average hop count of 0. When there are more than 3 users spread throughout the network, the 3 CE-Best cyber-entities cannot be near all the users simultaneously, resulting in an average hop count that is greater than 0. Figure 12 shows that the migration behavior of the CE2 cyber-entities allows them to adapt to the locations of the users in the network.

Although the scenarios used in the above simulations are specific to Aphid and may lack generality, we believe that these preliminary simulation results are very encouraging.

Our simulator is developed in Java. The current version of the simulator can simulate a wide variety of network topologies and user demand workloads. It can also simulate cyber-entities with different behavior policies. Simulator source code is available at netresearch.ics.uci.edu/bionet/resources/. The current simulator, however, lacks some key features such as the ability to dynamically add and remove nodes (to evaluate survivability of the Bio-Networking Architecture), to implement various mechanisms to create diversity in cyber-entity behaviors, and to dynamically create applications through interaction of multiple cyber-entities. Such features will be added to the simulator.

9   DISCOVERY IN THE BIO-NETWORKING ARCHITECTURE

In the Bio-Networking Architecture, a discovery mechanism provides means to locate distributed information, users, cyber-entities, and any group of cyber-entities that exist on the network. For instance, locating a unique object may occur when a user tries to locate another user on the network, or may occur when a cyber-entity tries to relocate or reconnect to a cyber-entity that has moved. Locating a group of objects or locating objects with certain attributes may occur when a user (or a cyber-entity) searches for an application (i.e., a group of cyber-entities) that matches certain parameters.

In designing a discovery mechanism suitable for the Bio-Networking Architecture, we utilize the discovery relationship among cyber-entities and model the phenomenon of *social networking* in human society. (Note that as described in section 4.2.1, cyber-entities collaborating in discovery may form a discovery relationship.) In human society, the proliferation of friend relationships among individuals has inspired the well-known "six degrees of separation" concept [60], which theorizes that any two persons in the world are separated by at most six relationships. In the Bio-Networking Architecture, cyber-entities are widely distributed, making contact with each other to create a large-scale connectivity of discovery relationships. Discovery in the Bio-Networking Architecture is conducted over this discovery relationship network in a manner similar to the social networking in human society (namely, a cyber-entity asks cyber-entities with which it has a relationship). In other words, search packets are forwarded to other cyber-entities along discovery relationship links in the proposed discovery mechanism. By using the discovery relationship network, searches adapt to dynamic aspects of a network. Many relationship pathways may exist for reaching a target cyber-entity or target cyber-entities, and thus, communication failure or cyber-entity loss is not likely to impact the reachability of cyber-entities during discovery, even when the discovery relationship network dynamically changes.

A detailed description of the proposed discovery mechanism follows [14].

9.1   KEY FEATURES OF THE PROPOSED DISCOVERY MECHANISM

To illustrate key features of the proposed discovery mechanism, let's consider the following scenario. A has a discovery relationship with B, and B also has a discovery relationship with C. A and C, however, do not have a direct discovery relationship between themselves. Assume that A asks B for information X. Assume also that B does not have information X, and B in turn asks C for X. C has information X and returns X to B. B then relays X to A. In this simple scenario, we make the following observations that form a basis for the proposed discovery mechanism.

1. Although B does not have information X and simply relays X from C to A, it appears to A as if B itself has information X.
2. B is not aware that A considers B as having information X.
3. If A needs to find X again, it is natural for A to ask B, as B was successful in returning X in the past. Note, however, that relying on the past may not always lead to a successful search, as the relationship between B and C may change, or a new cyber-entity may enter the discovery relationship and may return X quicker than B.
4. The identity of C (who returned a search hit) is hidden from A (who initiated the search), and vice versa. (Note that even B, an intermediate cyber-entity, does not generally know who returned a search hit or who initiated the search,

as A might have forwarded B a search packet from someone else, and similarly, C might have simply forwarded B a search hit from someone else.)

From observation (1), we argue that, in discovery, it is not critical to know who actually has the target information. Rather it is important to know in which direction the target information exists (or who knows who has the target information). Relying on the direction, not the exact location, of the target information makes the discovery mechanism robust to dynamic changes in the discovery relationship network. As long as a path to the target information exists, the discovery mechanism is likely to find it even when/after the target information moves. As will be described in section 9.2, the proposed discovery mechanism uses the community to remember the direction of the target information at cyber-entities.

From observation (2), we argue that it is the "asking" side of the discovery relationship (i.e., A in the above example), not the "asked" side of the discovery relationship (i.e., B in the above example), that maintains information on what the "asked" side knows. Since the "asking" side knows what information is being sought at the moment, and thus, it can only store who knows the information being sought. This reduces the amount of information that needs to be kept (compared to the "asked" side anticipating what information may be sought and storing all the keywords it knows about, regardless of whether they are actually needed in discovery or not). As described later, in the proposed discovery mechanism, community information (i.e., the information about which direction the target information exists) is stored at the "asking" side of the discovery relationship.

From observation (3), we argue that, when the discovery relationship network is dynamic, a search packet may be forwarded more (less) often to those who were more (less) useful in discovery in the past. Sending a search packet to those who were not very useful in the past makes the discovery mechanism robust to dynamic changes, as the past may not be a good indicator of future discovery performance when the discovery relationship network is dynamic. As will be described later in section 9.2, in the proposed discovery mechanism, a search originator cyber-entity expresses the degree of its satisfaction with a returned search hit, and the keyword strength of a cyber-entity (i.e., the measure of the usefulness of a cyber-entity in discovery) is adjusted based on the degree of satisfaction of a search originator cyber-entity.

From observation (4), we argue that relaying search packets and search hits preserves the anonymity of search originators and those who return search hits. As will be described in section 9.2, in the proposed discovery mechanism, search packets and search hits propagate over the relationship network hop-by-hop and are not exchanged directly between the search originator and those who return search hits.

The following section 9.2 describes how the above features are implemented in the proposed discovery mechanism.

9.2   COMMUNITY AND KEYWORD STRENGTH

In the proposed discovery mechanism, a discovery is initiated by a single entity (a user or a cyber-entity). Search packets contain the packet ID (globally unique ID given to a packet), packet type (indicating that the packet is for search), search ID (unique ID given to each search), and details (i.e., a set of keywords) to specify the attributes of the cyber-entity being sought (such as service type, age, and other attributes of a search target cyber-entity). In the following description, we assume that a search packet contains only one keyword, not multiple keywords, to describe a search target to simplify the explanation.

A search originator, upon receiving a search hit, examines the search hit, creates a *reward* message,[3] indicates the degree of its satisfaction with the received search hit in the *reward* message, and sends the *reward* message back to the cyber-entity who returned the search hit. A *reward* message back propagates the path that the search hit took to reach the search originator.

When a cyber-entity on the back-propagation path receives a *reward* message, it examines which keyword the *reward* message is for and to which neighboring cyber-entity this *reward* message should be forwarded (i.e., which neighboring cyber-entity is a downstream cyber-entity on the back propagation path). For the purpose of explanation, let's assume that cyber-entity A received a *reward* message that contains keyword X and that the downstream cyber-entity of the back propagation path is cyber-entity B (i.e., a *reward* message will be forwarded to B). Cyber-entity A examines whether its discovery relationship table has an entry for keyword X for cyber-entity B. If it does not, cyber-entity A creates an entry and stores keyword X with its initial keyword strength in its discovery relationship table to indicate that cyber-entity B has information regarding keyword X. If cyber-entity A already has an entry for keyword X for cyber-entity B, cyber-entity A adjusts the keyword strength of X based on the degree of satisfaction indicated in the *reward* message. Upon the arrival of a *reward* message, the cyber-entity who returned a search hit (containing keyword X) also adjusts the strength of keyword X based on the degree of satisfaction indicated in the *reward* message.

A simple policy for adjusting keyword strength based on the degree of the search originator's satisfaction, for example, may be to increment (decrement) the strength when the response from the search originator indicates positive (negative) satisfaction degree with the search hit. Thus, the strength of a keyword at a cyber-entity on a back propagation path indicates how useful the cyber-entity is for discovering a given keyword. The strength of a keyword at the cyber-entity who returned a search hit indicates how satisfied search originators are with the information it has.

---

[3]This *reward* message is used for discovery and is different from the reward message in the Bio-Networking ACL communication language described in section 4.3.

This keyword strength in discovery is distinct from the application strength used in the creation of applications (described in section 4.4), since only usefulness for searches is considered.

9.3   SEARCH FORWARDING PROCESS

Upon receiving a search packet, a cyber-entity first examines whether it has already received a search packet for the particular search, and if it already has, the search packet is discarded. (This is possible by storing at each cyber-entity for a certain time period a list of unique search ID's that it has seen.) If the cyber-entity has not yet seen the particular search, the cyber-entity decides to which cyber-entity it forwards the search packet. The policy for forwarding a search packet used in the proposed discovery mechanism is to choose a cyber-entity proportionally to its keyword strength (i.e., how useful a cyber-entity is in discovery). Note that, even if a relationship partner cyber-entity was not very useful in the past (indicated by a small keyword strength value), a search packet is forwarded with a small probability in the proposed algorithm. (This makes the proposed discovery algorithm robust to dynamic changes in the discovery relationship network.)

When one or more cyber-entities are selected as candidates for receiving search packets, they are forwarded to all the candidate cyber-entities, and at the same time, corresponding keyword strengths are reduced. This reduction in keyword strength in the search forwarding phase allows adaptive learning of the dynamically changing discovery relationship network as described below. If the current search leads to a successful search, the search originator will return a reward packet, and keyword strengths will be adjusted. If a search leads to a nonsuccessful search (for instance, due to changes in the location of the target cyber-entity or intermediate cyber-entities who relayed search packets or search hits), keyword strength stays at a reduced value. After a few nonsuccessful searches, keyword strength will be reduced to zero, at which point, corresponding entries are eliminated from the discovery relationship table to keep the discovery relationship table from growing large.

If the cyber-entity itself (who received a search packet) is selected as a candidate for forwarding a search packet, the cyber-entity generates a search hit, and the search hit is returned to the cyber-entity from whom a search packet came. The search hit back propagates the path that the search packet took. (See the description below in section 9.4 for details of the back propagation process.)

Keyword strength-based forwarding improves the efficiency of discovery. Since the strength represents the search-related usefulness of the cyber-entity, it is more likely that a search can be resolved by following the typically more useful relationship.

9.4 SEARCH BACK-PROPAGATION PROCESS

In the proposed discovery mechanism, search hits propagate back to the search originator following the same forwarding path that led to the target cyber-entity. This allows the search originator to remain anonymous to each target cyber-entity. In order to store the back propagation path information, each cyber-entity, during the forwarding phase, keeps a record of the unique search packet ID, the cyber-entity to which search hits should be sent (e.g., the cyber-entity one hop back toward the search originator), and to which cyber-entities the search packet was forwarded. The cyber-entity listed as an entity to send a search hit back need not be the cyber-entity immediately before the current cyber-entity and may be a cyber-entity several hops back along the path toward the search originator. This ability to select which cyber-entities will handle the responses during the backtracking phase can help balance the load, since some cyber-entities may have more available resources for storing and processing search responses.

If discovery generates multiple search hits, the back propagation process can include filtering of search hits at each hop back toward the search originator. This filtering ensures that the maximum number of search hits that any single cyber-entity must deal with remains reasonably small. Since not all search hits match the search criteria equally, search hits may be evaluated against the search criteria, and poor search hits can be removed along the path back to the search originator.

9.5 COMPARISON TO EXISTING PEER-TO-PEER DISCOVERY MECHANISMS

Many existing discovery algorithms for peer-to-peer networks are based on broadcasting search packets over the discovery relationship network. A search packet is broadcast with a time to live (TTL) value to limit the scope of search and to reduce high overhead associated with broadcasting. Although the use of TTL reduces the amount of overhead, it is difficult to determine an optimal TTL value for a given network, and thus, the scalability of existing discovery algorithms may be limited. In addition, in existing discovery algorithms, all search target entities that match the search keywords return search hits, independently of whether they have been considered useful in the past by users (i.e., query originators), potentially leading to a large number of non-useful search hits returning to a user.

In the proposed discovery mechanism, each cyber-entity gradually acquires information on its community (i.e., information on what neighboring cyber-entities know), and keyword strengths to neighboring cyber-entities are adjusted based on users' levels of satisfaction. Communities and keyword strengths are used in determining where a search packet may be forwarded next and whether a cyber-entity that received a search packet should return a query hit or not.

Since search packet forwarding is done probabilistically, there is no need to specify the TTL value in the proposed discovery mechanism. In the proposed discovery mechanism, when a search packet arrives at a cyber-entity whose keyword strength is large, a search packet is likely to be forwarded to the next hop cyber-entity, pursuing a promising discovery path. When a search packet arrives at a cyber-entity whose keyword strength is small, a search packet is likely to be discarded, terminating the unpromising search path. In addition, since a search packet is likely to be forwarded (probabilistically) to those cyber-entities with higher keyword strength, unnecessary forwarding of search packets is avoided in the proposed discovery mechanism. Finally, community and probabilistic forwarding makes the proposed discovery mechanism robust to changes in the discovery relationship network.

10   STABILITY ANALYSIS OF THE BIO-NETWORKING ARCHITECTURE

We have investigated whether the Bio-Networking Architecture operates at an equilibrium point. A mathematical model was built, and conditions in which the Bio-Networking Architecture is stable were obtained [46]. In the mathematical model analyzed, a cyber-entity provides a service to users in exchange for energy. It pays energy to the platform software for using the resource that the platform software manages. A utility function is associated with the platform software, and the platform software determines the prices of resources based on the utility function. Similarly, a utility function is associated with a cyber-entity, and a cyber-entity determines the amount of resources it uses based on its utility function. Utility for a cyber-entity, and thus, the amount of resource that a cyber-entity uses, depends on various system parameters (such as the amount of resource used by other nearby cyber-entities, and the prices of the resources and the number of users that are nearby). Similarly, utility of the platform software depends on various system parameters such as resource prices on other platforms and the amount of resources that cyber-entities use. A mathematical model was created and analyzed to obtain conditions in which the model is stable.

11   CURRENT AND FUTURE WORK ON THE BIO-NETWORKING ARCHITECTURE

In this section we briefly describe some of the key research questions we are currently investigating in the Bio-Networking Architecture project.

- *Which concepts from the biological world can provide new and beneficial approaches to the design of network applications?* Some researchers have identified the immune system and self/non-self discrimination as concepts which could benefit the design of computer security and intrusion detection systems [12, 15, 37]. We have also identified many concepts (described in sections

2 and 6) as concepts that can improve the network. We plan to investigate a spectrum of biological concepts and mechanisms that can be incorporated in the Bio-Networking Architecture.

- *What is the overhead of the biological concepts and mechanisms implemented in the Bio-Networking Architecture, and how does it impact the scalability of the Bio-Networking Architecture?* We plan to evaluate the overhead of biological concepts and mechanisms implemented in the Bio-Networking Architecture and their impact on scalability and seek efficient implementation techniques for these mechanisms. We believe that the overhead will be acceptable given efficient implementations and the rapidly increasing storage and communication capacity of future networking environments.

- *What is the relationship between the individual cyber-entity behaviors and their emergent behavior? To what extent is it possible to predict or control the emergent behaviors so that only the desired effects are produced?* In the Bio-Networking Architecture, a vast number of cyber-entities with different functionality and different behavioral patterns will be interacting. We believe that by carefully defining and designing the individual cyber-entity behaviors, the desired emergent behaviors (such as adaptation, evolution, security, and survivability) will emerge. The relationship between the diverse cyber-entity behaviors and their emergent behaviors will be investigated and characterized.

- *What is a suitable mechanism to implement cyber-entity behaviors and what is an efficient mechanism to create diversity in cyber-entity behaviors?* In the current design, each cyber-entity behavior is implemented using factors and weights. Mutation and crossover in replication and reproduction are done to behavior factors and weights. We plan to evaluate the efficiency of these diversity creation mechanisms, and if they are not efficient, we plan to seek other mechanisms to implement cyber-entity behaviors and to create behavioral diversity in cyber-entities.

- *Can applications built using the Bio-Networking Architecture successfully evolve to more suitable behaviors?* Applications built using the Bio-Networking Architecture need to progress through enough generations in a relatively short period of time so that evolutionary improvements are perceived by the users. We plan to characterize the time needed for perceivable evolutionary improvements using the simulation and test network environments. (Note that the Bio-Networking Architecture does not solely rely on evolution to produce suitable behaviors. Cyber-entities can be allowed to evolve initially in an experimental or simulated environment. Thus, when they are introduced into the network, their behaviors will perform reasonably well.)

- *How does the dynamic nature of the Bio-Networking Architecture impact its performance and scalability?* Many properties of the Bio-Networking Architecture are dynamic. Cyber-entities may move frequently; new cyber-entities are created frequently through replication and reproduction; many cyber-entities die due to lack of energy or old age. These dynamic properties may impact the performance of the Bio-Networking Architecture.

- *How do we protect network nodes from malicious cyber-entities and cyber-entities from each other and malicious network nodes?* Network nodes in the Bio-Networking Architecture have a software layer (i.e., platform software) that controls resource access and protects the node from malicious cyber-entities. However, the problem of protecting cyber-entities from malicious nodes is still unresolved. This is a general problem in mobile agent systems and active networks, and this security issue needs to be fully investigated.

- *Does the Bio-Networking Architecture exhibit correct behaviors? Are the applications provided on the Bio-Networking Architecture semantically correct, and do cyber-entities exhibit semantically correct behaviors?* In the Bio-Networking Architecture, cyber-entities autonomously exchange information with other cyber-entities and collectively provide applications. However, the ability of multiple cyber-entities to exchange information may not guarantee that they can collectively provide an application that is semantically correct. Mechanisms need to be investigated and incorporated into the Bio-Networking Architecture design to check whether applications created on the Bio-Networking Architecture are semantically correct and meet users' expectations. In addition, in the Bio-Networking Architecture, the evolution mechanisms (i.e., mutation and crossover) create new cyber-entity behavior algorithms. Similarly, mechanisms need to be investigated and incorporated into the Bio-Networking Architecture design to check whether cyber-entity behavior algorithms created through the evolution mechanisms are semantically correct. These correctness issues (correctness of applications and cyber-entity behaviors) need to be investigated.

We plan to employ simulation and analysis approaches to investigate the key research questions explained above. As described in section 8, we have conducted a feasibility study of the Bio-Networking Architecture through simulations. Initial results show that the Bio-Networking Architecture exhibits such key features as adaptability, survivability and availability. However, the current simulation study does not examine some of the key characteristics of the Bio-Networking Architecture. For instance, the evolution and natural selection capability of the Bio-Networking Architecture was not examined in the simulation study presented in section 8. The scalability of the simulation configuration was also limited, and the range of biological mechanisms and concepts simulated was also limited in section 8. The dynamic creation of applications through the interaction of multiple cyber-entities was not investigated in the current simulations. We plan to evaluate the Bio-Networking Architecture through more extensive and thorough simulations and analysis.

In addition, we plan to evaluate the Bio-Networking Architecture empirically through implementation and deployment. Components of the Bio-Networking Architecture (e.g., platform software, cyber-entities, dynamic service creation mechanisms, discovery mechanisms) will be designed and implemented, and the Bio-Networking Architecture will be deployed initially on a small, isolated test network, and later on bigger testbeds such as vBNS, CalREN2, and the Internet. We plan to demonstrate that the Bio-Networking Architecture is capable of generating the desired emergent behaviors such as adaptation, evolution, security, and survivability, through empirical evaluation of the deployed Bio-Networking Architecture. The rate of adaptation and evolution, as well as the security and survivability, of the Bio-Networking Architecture will be monitored. The benefits and limitations of the Bio-Networking Architecture will be empirically evaluated.

12 RELATED WORK

The Bio-Networking Architecture project touches on a wide variety of areas, and related work is described below.

12.1 EMERGENT SYSTEMS

There have been numerous works which simulate the emergent behavior and evolution of biological architectures [4, 11, 13, 38, 39, 41, 69]. However, these works are confined to the simulation of biological architectures and comparisons of the results to what is observed in nature. Some researchers have applied the concept of cellular programming (parallel, asynchronous evolutionary algorithms) to build evolving hardware capable of basic computational tasks [20, 59]. The Bio-Networking Architecture project goes one step further by applying biological concepts to large-scale complex network services and applications.

Multi-agent systems and artificial life systems simulate biological processes to cooperatively solve distributed problems. Results in the area of multi-agent systems and artificial life have shown that useful emergent behavior, such as ant-foraging [6] or arranging blocks into rows [71], can arise from only local interaction. Such results underscore the potential success of the Bio-Networking Architecture.

12.2 BIOLOGICALLY INSPIRED INTRUSION DETECTION SYSTEMS

There have been some works which apply immunological principles to intrusion detection and anti-(computer) virus applications [12, 15, 37]. D'haeseleer et al. [12] generate bit patterns that only match abnormal bit patterns in application binaries. Forrest et al. [15] monitor system call patterns for privileged UNIX programs. If the system call patterns deviate from typical patterns stored in a database, an alarm is triggered. In Kephart [37], viruses are lured into infecting decoy system programs to automatically construct a database of known computer viruses. The Bio-Networking Architecture is more general than the above approaches because it attempts to apply general biological principles to all aspects of networking. (The above approaches only apply immunological principles to intrusion or virus detection.) Also, in the Bio-Networking Architecture, intrusion detection can be implemented by the application components themselves, i.e., by the cyber-entities. This eliminates the need for external programs or a pattern database. There has been other work on intrusion detection systems that are not based on biological models [3, 23, 24, 54, 78].

12.3 DYNAMIC SERVICE COMPOSITION

Currently, some frameworks and architectures exist for dynamically creating services. One such example is Hive [44], where a service is provided through the interaction of distributed agents. In Hive, agents interact by specifying the Java interface object of the interaction partner. Thus, interaction in Hive is limited to agents that mutually implement the interface object of the partner. Unlike Hive, the Bio-Networking Architecture supports message-based communication for cyber-entity interactions, and thus, a cyber-entity in the Bio-Networking Architecture can communicate with other cyber-entities without implementing a special interface for a particular cyber-entity. Further, in the Bio-Networking architecture, there is no centralized entity to manage cyber-entities, and thus, it scales in the number of cyber-entities, while Hive agents store their attributes in a centralized directory, which can be a scalability bottleneck. Bee-gent [36] is another example of a framework for creating services dynamically. It uses a centralized interaction protocol model; a mediator agent maintains a centralized IP and coordinates agent interactions to reduce complexity in IPs. With this centralized IP approach, Bee-gent restricts the flexibility and scalability of the agent cooperation. Unlike Bee-gent, no centralized IP is used in the Bio-Networking Architecture.

12.4 GENETIC ALGORITHMS

Conventional genetic algorithms (GAs) [42, 45] evaluate all members of the population at distinct time intervals and allow the best members to reproduce. Although the Bio-Networking Architecture borrows some concepts from genetic algorithms, it takes a significantly different approach. The Bio-Networking Architecture uses GA mechanisms (e.g., mutation and crossover) to introduce algorithmic diversity into the cyber-entity population. However, because the Bio-Networking Architecture does not have centralized control and works in a large network environment, it does not use global evaluation of the cyber-entity population as in conventional GAs.

12.5 DISCOVERY IN PEER-TO-PEER SYSTEMS

A number of peer-to-peer systems exist. Many existing peer-to-peer systems are based on a centralized file sharing architecture. For instance, in Napster [50], Xdegree [81], PeerGenius [53], and ChainCast [7], a centralized directory maintains information on shared files, and network nodes search for a file on the centralized directory. Unlike these systems, discovery in the Bio-Networking Architecture is distributed. Existing peer-to-peer systems that are based on distributed file sharing architecture include Freenet [9], Gnutella [19], Jnutella [34], Ohaha [51], Google [52], and IBM's Clever System [8]. In Freenet, Gnutella, Jnutella, search packets are transmitted between nodes following pointers. Unlike the Bio-Networking Architecture, Freenet does not use the relationship strength (i.e., usefulness of a node) in search. Google and Clever System (search engines) use the node ranking, usefulness of a node, in search. The node ranking is calculated based on the number of links to the node and where the links are from. The Bio-Networking Architecture also uses relationship strength in search, however, search relationship strength is more general and may support similar and/or other factors. The Bio-Networking Architecture also tries to use relationship clustering to improve efficiency. Other notable peer-to-peer systems include JXTA [21] (which does not implement any specific discovery mechanism, but provides a set of primitives that a user can use to create his own discovery) and Mojo Nation [47] (a file sharing system that uses mojo, which corresponds to energy in the Bio-Networking Architecture, for hosting someone else's files).

Discovery through cyber-entities is quite different in that the files (represented as cyber-entities) themselves contain relationships. Discovery is decentralized and progresses along these relationships and not along platform connections. The relationships are also adaptive and may support many different styles of organization. With proper organization, many different types of searches should perform relatively efficiently, whereas the previously described discovery mechanisms are aimed at propagating searches based only on a single style of organization.

12.6   MOBILE AGENT SYSTEMS

Current popular mobile agent systems, including IBM's Aglets [40], Object Space's Voyager [70], General Magic's Odyssey [18], and the University of Stuggart's Mole project [48], adopt the view that a mobile agent is a single unit of computation. They do not employ biological concepts and massive replication as fundamental mechanisms, nor take the view that a group of agents may be viewed as a single functioning collective entity (i.e., a group of cyber-entities providing an application in the Bio-Networking Architecture). They also allow mobile agents to use unlimited CPU and memory resources at the nodes they visit [35]. The Bio-Networking Architecture limits the number of resources permitted for use by a cyber-entity in proportion to the amount of energy each cyber-entity pays, thus creating a free market of computing resources for cyber-entities.

13 CONCLUSIONS

We envision in the future that the Internet will span the entire globe, interconnecting all humans and all man-made devices and objects. When a network scales to this magnitude, it will be virtually impossible to manage a network through a central, coordinating entity. A network must be autonomous and contain built-in mechanisms to support such key features as scalability, adaptability, simplicity, and survivability. We believe that applying concepts and mechanisms from the biological world provides a unique and promising approach to solving key issues that future networks face.

The Bio-Networking Architecture that we presented in this chapter is based on key biological concepts and mechanisms. Designs, as well as major system features and characteristics of the Bio-Networking Architecture were described in this chapter. The Bio-Networking Architecture has attracted a number of researchers outside of the University of California, Irvine. NTT, for instance, has started investigating the Ja-Net (Jack-in-the-Net) Architecture inspired by and based on the Bio-Networking Architecture. The work presented in subsections 4.3, 4.4, and 4.5 of this chapter are jointly developed by the Bio-Networking Architecture project researchers at the University of California, Irvine and the Ja-Net Architecture researchers at NTT. The work presented in other sections is developed by the Bio-Networking Architecture project researchers at the University of California, Irvine.

With the Bio Networking Architecture proposed in this chapter, each and every user of the future global Internet implements his or her own services in a cyber-entity and releases it into the Internet. There will be millions and millions of cyber-entities (created by users throughout the world) on the Internet, autonomously migrating, replicating, reproducing, and establishing relationships with other cyber-entities that provide useful services. Diverse applications, as well as diverse cyber-entity behavior algorithms, emerge through autonomous interactions of cyber-entities. Cyber-entities that provide a popular service and possess fit behaviors survive and prosper, and those with unpopular services and unfit behavior algorithms perish through natural selection. The future Internet will be autonomous and live its own life without human intervention.

ACKNOWLEDGMENTS

Work by Tatsuya Suda is supported by the National Science Foundation through grants ANI-0083074 and ANI-9903427, by DARPA through Grant MDA972-99-1-0007, by AFOSR through Grant MURI F49620-00-1-0330, and by grants from the University of California MICRO and CoRe programs, Program, Hitachi, Hitachi America, Novell, Nippon Telegraph and Telephone Corporation (NTT), NTT Docomo, Fujitsu, Denso, and Nippon Steel Information and Communication Systems Incorporated (ENICOM).

Tatsuya Suda also holds the title of NTT Research Professor, and his NTT contact information is the same as the co-authors' contact information.

REFERENCES

[1] Austin, J. L. *How to Do Things with Words*. Cambridge, MA: Harvard Uniersity Press, 1962.

[2] Baas, N. "Emergence, Hierarchies, and Hyperstructures." In *Artificial Life III*, edited by Christopher Langton, 515–538. Reading, MA: Addison-Wesley, 1994.

[3] Balasubramaniyan, J., J. Fernandez, D. Isacoff, E. Spafford, and D. Zamboni. "An Architecture for Intrusion Detection Using Autonomous Agents." Technical Report, Purdue Coast Laboratory, Purdue University, West Lafayette, IN, 1998.

[4] Boekhorst, I., and P. Hogeweg. "Effects of Tree Size on Travelband Formation in Orang-Utans." In *Artificial Life IV: Proceedings of the Fourth International Workshop on the Synthesis and Simulation of Living Systems*, edited by Rodney Brooks and Pattie Maes. Cambridge, MA: MIT Press, 1994.

[5] Branden, C., and J. Tooze. *Introduction to Protein Structure*. New York, NY: Garland Publishing Inc., 1991.

[6] Bruckstein, A. "Why the Ant Trails Look So Straight and Nice." *The Mathematical Intelligencer* **15(2)** (1993): 59–62.

[7] ChainCast. Home Page. 2004. ⟨http://www.chaincast.com/⟩.

[8] Chakrabarti, S., B. Dom, D. Gibson, J. Kleinberg, S. R. Kumar, P. Raghavan, S. Rajagopalan, and A. Tomkins. "Hypersearching the Web." *Sci. Am.* **2** (1999): 54–60.

[9] Clarke, I., O. Sandberg, B. Wiley, and T. W. Hong. "Freenet: A Distributed Anonymous Information Storage and Retrieval System in Designing Privacy Enhancing Technologies." In *Proc. of the International Workshop on Design Issues in Anonymity and Unobservability*, LNCS 2009. Springer, 2001.

[10] Cockayne, W., and M. Zyda. *Mobile Agents*. Greenwich, CT: Manning/Prentice Hall, 1997.

[11] Collins, R., and D. Jefferson. "AntFarm: Towards Simulated Evolution." In *Artificial Life II*, edited by C. G. Langton, C. Taylor, J. D. Farmer, and S. Rasmussen, 579–602. Reading, MA: Addison-Wesley, 1992.

[12] D'haeseleer, P., S. Forrest, and P. Helman. "An Immunological Approach to Change Detection." In *Proc. of the IEEE Symposium on Security and Privacy*, 110–119. Washington, DC: IEEE Press, 1996.

[13] Eggenberger, P., and R. Dravid. "An Evolutionary Approach to Pattern Formation Mechanisms on Lepidopteran Wings." In *Proc. of the Congress on Evolutionary Computation-CEC '99*, vol. 1, 366–373. Washington, DC: IEEE Press, 1999.

[14] Enomoto, A. "Community Based Discovery in Peer-to-Peer Networks." Master's thesis, Kyoto University, 2002.

[15] Forrest, S., S. Hofmeyr, A. Somayaji, and T. Longstaff. "A Sense of Self for Unix Processes." In *Proc. of the IEEE Symposium on Security and Privacy*, 120–128. Washington, DC: IEEE Press, 1996.

[16] Foundation for Intelligent Physical Agents, FIPA Communicative Act Library Specification. 2000. ⟨http://www.fipa.org⟩.

[17] Franks, N. "Army Ants: A Collective Intelligence." *Amer. Sci.* **77** (1989): 139–145.

[18] General Magic Inc. Odyssey Home Page. 1997.
⟨http://www.genmagic.com/technology/odyssey.html⟩.

[19] Gnutella. Home Page. ⟨http://gnutella.wego.com/⟩.

[20] Goeke, M., M. Sipper, D. Mange, A. Stauffer, E. Sanchez, and M. Tomassini. "Online Autonomous Evolware." *Proc. of the First International Conference on Evolvable Systems: from Biology to Hardware (ICES96)*, 96–106, LCNS 1259. Springer, 1997.

[21] Gong, Li. "JXTA: A Network Programming Environment." *IEEE Internet Computing* **5(3)** (2001): 88–95.

[22] Gotwald, W. *Army Ants.* Ithaca, NY: Comstock Publishing Associates, Cornell University Press, 1995.

[23] Heberlein, L., G. Dias, K. Levitt, B. Mukherjee, J. Wood, and D. Wolber. "A Network Security Monitor." *Proc. of the IEEE Symposium on Security and Privacy*, 296–304. Washington, DC: IEEE Press, 1990.

[24] Hochberg, J., K. Jackson, C. Stallings, J. McClary, D. DuBois, and J. Ford. "NADIR: An Automated System for Detecting Network Intrusion and Misuse." *Comp. & Security* **12(3)** (1993): 235–248.

[25] Iizuka, T., Angel Lau, and T. Suda. "A Design of Local Resource Access Control for Mobile Agent in PDA." In *Proceedings of the Asian-Pacific Conference on Communications.* 2001. ⟨http://netresearch.ics.uci.edu/bionet/publications/⟩.

[26] Imada, M., Y. Katayama, M. Matsuo, and T. Suda. "Service Creation Based on Service Attributes." In *Proceedings of IEICEJ Assurance Symposium* (written in Japanese). 2001. ⟨http://netresearch.ics.uci.edu/bionet/publications/⟩.

[27] International Business Machines. Aglets Home Page. March 2002. ⟨http://www.trl.ibm.com/aglets/⟩.

[28] International Business Machines Corporation. "Mobile Agent Facility Specification." OMG TC Document cf/96-08-01, 1996.

[29] Itao, T., T. Suda, T. Nakamura, and M. Matsuo. "Adaptive Networking Architecture for Service Emergence." In the *Proceedings of the 2001 Symposium on Applications and the Internet (SAINT 2001).* Online publication. IEEE Computer Society. 2001. ⟨http://computer.org/proceedings/saint/0942/0942toc.htm⟩.

[30] Itao, T., T. Suda, and T. Aoyama. "Jack-in-the-Net: Adaptive Networking Architecture for Service Emergence." *Proceedings of the 7th Asia-Pacific Conference on Communications (APCC2001). IEICE Transactions on Communications* (2001).

[31] Itao, T., T. Suda, T. Nakamura, M. Imada, and M. Matsuo. "Jack-in-the-Net: Adaptive Networking Architecture for Service Emergence." Invited Paper. In *Trans. Inst. Electronics Commun. Engineers of Japan (IECEJ)*, Vol.J84-B, No.3, 310–320. 2001.

[32] Itao, T., T. Nakamura, M. Matsuo, T. Suda, and T. Aoyama. "The Model and Design of Cooperative Interaction for Service Composition." *The DI-COMO* (2001): submitted.

[33] Itao, T., T. Nakamura, M. Matsuo, T. Suda, and T. Aoyama. "Service Emergence based on Cooperative Interaction of Self-Organizing Entities." In the Proceedings of the 2001 Symposium on Applications and the Internet (SAINT 2001). Online publication. IEEE Computer Society. 2001. ⟨http://computer.org/proceedings/saint/0942/0942toc.htm⟩.

[34] Jnutella Peer-to-Peer Protocol (JPPP) Specification (in Japanese). ⟨http://www.jnutella.org/⟩.

[35] Karjoth, G., D. Lange, and M. Oshima. "A Security Model For Aglets." *IEEE Internet Computing* **1(4)** (1997): 68–77.

[36] Kawamura, T., Y. Tahara, T. Hasegawa, A. Ohsuga, and S. Honiden. "Bee-Agent: Bonding and Encapsulation Enhancement Agent Framework for Development of Distributed Systems." *J. IEICE* **J82-D-I(9)** 1999.

[37] Kephart, J. "A Biologically Inspired Immune System for Computers." In *Artificial Life IV: Proceedings of the Fourth International Workshop on the Synthesis and Simulation of Living Systems*, edited by Rodney Brooks and Pattie Maes. Cambridge, MA: MIT Press, 1994.

[38] Koza, J. "Evolution of Emergent Cooperative Behavior Using Genetic Programming." *Computing with Biological Metaphors*. London, UK: Chapman & Hall, 1994.

[39] Kyoda, K., and H. Kitano. "Simulation of Genetic Interaction for Drosophila Leg Formation." *Proc. of the Pacific Symposium on Biocomputing*, 77–89. Online proceedings. 1999. ⟨http://www-smi.stanford.edu/projects/helix/psb99/⟩.

[40] Lange, Danny B., and Mitsuru Oshima. *Programming & Deploying Mobile Agents with Java Aglets*. Reading, MA: Addison-Wesley, 1998.

[41] Luthi, P. O., B. Chopard, A. Preiss, and J. J. Ramsden. "A Cellular Automaton Model for Neurogenesis in Drosophila." *Physica D* **118(1-2)** (1998): 151–160.

[42] Michalewicz, Z. *Genetic Algorithms + Data Structures = Evolution Programs*. Springer-Verlag, 1996.

[43] Michener, C. *The Social Behavior of the Bees*. Cambridge, MA: The Belknap Press, 1974.

[44] Minar, N., M. Gray, O. Roup, R. Krikorian, and P. Maes. "Hive: Distributed Agents for Networking Things." *Proceedings of the First International Symposium on Agent Systems and Applications (ASA'99)/Third International Symposium on Mobile Agents (MA'99)*, 118–129. Washington, DC: IEEE Press, 1999.

[45] Mitchell, M. *An Introduction to Genetic Algorithms.* Cambridge, MA: MIT Press, 1996.

[46] Miyamoto, N. "An Equilibrium Model of a Self-Organizing Network Architecture." Master's thesis, Kyoto University, 2001.

[47] Mojo Nation Technology Overview. ⟨http://www.mojonation.com/⟩.

[48] Mole Project Home Page. ⟨http://inf.informatik.uni-stuttgart.de/ipvr/vs/projekte/mole.html⟩.

[49] Nakamura, T., M. Matsuo, T. Itao, and T. Suda. "System Control in Ja-Net, a New Network Architecture with Service Emergence." IEICEJ Next Generation Networks workshop, December, 2001

[50] Napster. Home Page. 2004. ⟨http://www.napster.com/⟩.

[51] The Ohaha Systems. No longer available as of June 2001. "Smart Decentralized Peer-to-Peer Sharing Taking Gnutella, Freenet to the Next Level." ⟨http://www.ohaha.com/design.html⟩.

[52] Page, L., S. Brin, R. Motwani, and Terry Winograd. "The PageRank Citation Ranking: Bringing Order to the Web." Technical Report, Stanford University, Stanford, CA, 1998.

[53] PeerGenius.com. Home Page. 2004. ⟨http://www.peergenius.com/⟩.

[54] Porras, P., and P. Neumann. "EMERALD: Event Monitoring Enabling Responses to Anomalous Live Disturbances." *Proc. of the 20th National Information Systems Security Conference*, 353–365. NIST/NCSC, 1997.

[55] Roberts, S. J., W. Penny, and I. Rezek. "Temporal and Spatial Complexity Measures for EEG-based Brain-Computer Interfacing." *Med. & Biol. Eng. & Comp.* **37(1)** (1998): 93–99.

[56] Searle, J. R. *Speech Acts.* Cambridge, MA: Cambridge University Press, 1969.

[57] Seeley, T. "The Honey Bee Colony as a SuperOrganism." *Amer. Sci.* **77** (1989): 546–553.

[58] Seeley, T. *The Wisdom of the Hive.* Cambridge, MA: Harvard University Press, 1995.

[59] Sipper, M., M. Goeke, D. Mange, A. Stauffer, E. Sanchez, and M. Tomassini. "The Firefly Machine: Online Evolware." *Proc. of the IEEE Fourth International Conference on Evolutionary Computation (ICEC '97)*, 181–186. Washington, DC: IEEE Press, 1997.

[60] Six Degrees. Home Page. 2004. ⟨http://www.sixdegrees.com⟩.

[61] Song, S., and T. Suda. "Security Considerations in the Bio-Networking Architecture." *Proc. of the JSPS Workshop on Applied Information Technology for Science*, January 2001. Available on line. ⟨http://netresearch.ics.uci.edu/bionet/publications/⟩.

[62] Song, S., and T. Suda. "Security on Energy Level in the Bio-Networking Architecture." *Proc. of the 3rd International Conference on Advanced Communications Technology*, 2001. Available on line. ⟨http://netresearch.ics.uci.edu/bionet/publications/⟩.

[63] Steels, L. "Towards a Theory of Emergent Functionality." *Animals to Animats* **1** (1990): 451–461.

[64] Suda, T. "Jack-in-the-Net (Ja-Net) Architecture for Service Emergence." Proceedings of IEICEJ Assurance Symposium (written in Japanese). 2001. ⟨http://netresearch.ics.uci.edu/bionet/publications/⟩.

[65] Suda, T., T. Itao, T. Nakamura, and M. Matsuo. "Adaptive Networking Architecture for Service Emergence." *J. IECEJ* **J84-B(3)** (2001): 310–320.

[66] Suzuki, J., and T. Suda. "Bio Net Platform Design." 2002. ⟨http://netresearch.ics.uci.edu/bionet/publications/suzuki_jwaits01.ppt⟩

[67] Suzuki, J., and T. Suda. "Bionet Project Overview: Applying Biological Concepts and Mechanisms for Designing Adaptable, Scalable and Survivable Communication Software." *Proc. of the JSPS Workshop on Applied Information Technology for Science*. 2001. Available on line. ⟨http://netresearch.ics.uci.edu/bionet/publications/⟩.

[68] Tonegawa, S. "Somatic Generation of Antibody Diversity." *Nature* **302(5909)** (1983): 575–581.

[69] Toquenaga, Y., I. Kajitani, and T. Hoshino. "Egrets of a Feather Flock Together." In *Artificial Life IV: Proceedings of the Fourth International Workshop on the Synthesis and Simulation of Living Systems.*, edited by Rodney Brooks and Pattie Maes. Cambridge, MA: MIT Press, 1994.

[70] Voyager. Home Page. Information may be available at ⟨http://www.recursionsw.com/voyager.htm⟩.

[71] Wagner, I., and A. Bruckstein. "Row Straightening Via Local Interactions." Technical Report CIS-9406, Center for Intelligent Systems, Technion, Haifa, May 1994.

[72] Wang, M., and T. Suda. "The Bio-Networking Architecture." Proceedings, Gigabit Networking Workshop GBN99, New York, NY, March 21, 1999. Washington, DC: IEEE, 1999. ⟨http://www.comsoc.org/tcgn/conference/gbn99/⟩.

[73] Wang, M., and T. Suda. "The Bio-Networking Architecture: A Biologically Inspired Approach to the Design of Scalable, Adaptive, and Survivable/Available Network Applications." Technical Report 00-03, Department of Information and Computer Science, University of California, Irvine, 2000.

[74] Wang, M., and T. Suda. "The Bio-Networking Architecture: A Biologically Inspired Approach to the Design of Scalable, Adaptive, and Survivable/Available Network Applications." Proceedings of the IEEE Symposium on Applications and the Internet (SAINT), 43. Washington, DC: IEEE Press, 2001. ⟨http://netresearch.ics.uci.edu/bionet/publications/⟩.

[75] Wang, M., and T. Suda. "The Bio-Networking Architecture and Secure and Survivable Network Services and Applications." Paper presented at

the Cross-Industry Working Team Workshop, Reston, VA, July 17, 1998. ⟨http://www.xiwt.org/documents/July98Wkshp.html#Minutes⟩.

[76] Wang, M., and T. Suda. "Overview of the Bio-Networking Architecture." Paper presented at the New Research Session, ACM SIGCOMM Symposium, Sept. 2, 1999. ⟨http://netresearch.ics.uci.edu/bionet/publications/⟩.

[77] Wasserman, E. "The Network Is the Solar System: Vint Cerf on an Interplanetary Internet." *The Industry Standard* July 22, 1998.

[78] White, G., E. Fisch, and U. Pooch. "Cooperating Security Managers: A Peer-Based Intrusion Detection System." *IEEE Network* **10(1)** (1994w): 20–23.

[79] Wolpert, L., R. Beddington, J. Brockes, T. Jessell, P. Lawrence, and E. Mayerowitz. *Principles of Development.* New York: Oxford University Press, 1998.

[80] World Wide Web Consortium (W3C). Extensible Markup Language (XML) 1.0 (Second Edition). October 2000. ⟨http://www.w3.org/TR/2000/REC-xml-20001006⟩.

[81] Xdegrees. Home Page. 2004. ⟨http://www.xdegrees.com/⟩.

Index

Printed in the United States
By Bookmasters